AF327146

Ernst Schering Research Foundation
Workshop 56

Cytokines as Potential Therapeutic Targets for Inflammatory Skin Diseases

R. Numerof, C.A. Dinarello, K. Asadullah
Editors

With 30 Figures

Springer

Series Editors: G. Stock and M. Lessl

Library of Congress Control Number: 2005925477

ISSN 0947-6075

ISBN-10 3-540-25427-7 Springer Berlin Heidelberg New York
ISBN-13 978-3-540-25427-0 Springer Berlin Heidelberg New York

This work is subject to copyright. All rights are reserved, whether the whole or part of the material is concerned, specifically the rights of translation, reprinting, reuse of illustrations, recitation, broadcasting, reproduction on microfilms or in any other way, and storage in data banks. Duplication of this publication or parts thereof is permitted only under the provisions of the German Copyright Law of September 9, 1965, in its current version, and permission for use must always be obtained from Springer-Verlag. Violations are liable for prosecution under the German Copyright Law.

Springer is a part of Springer Science+Business Media
springeronline.com

© Springer-Verlag Berlin Heidelberg 2006
Printed in Germany

The use of general descriptive names, registered names, trademarks, etc. in this publication does not emply, even in the absence of a specific statemant, that such names are exempt from the relevant protective laws and regulations and therefor free for general use. Product liability: The publisher cannot guarantee the accuracy any information about dosage and application contained in this book. In every induvidual case the user must check such information by consulting the relevant literature.

Editor: Dr. Ute Heilmann, Heidelberg
Desk Editor: Wilma McHugh, Heidelberg
Production Editor: Michael Hübert, Leipzig
Cover design: design & production, Heidelberg
Typesetting and production: LE-TEX Jelonek, Schmidt & Vöckler GbR, Leipzig
21/3152/YL – 5 4 3 2 1 0   Printed on acid-free paper

Preface

Cytokines have been a focus of scientific interest at least since the 1980's. Analyzing their expression patterns has permitted a better understanding of the pathogenesis of various diseases, including the inflammatory dermatoses. Exploring T-cell mediated skin diseases such as psoriasis, atopic dermatitis, and allergic contact dermatitis in this context is of particular interest since the skin is becoming increasingly recognized as a "visible immunological organ". These T-cell mediated diseases are characterized by abnormal cytokine expression patterns and can be considered as "model diseases" for other inflammatory diseases such as rheumatoid arthritis, multiple sclerosis, organ transplant rejection, and inflammatory bowel disease.

Cytokines are now far beyond the stage when they were of interest only to the research sector as some cytokine-directed therapies are already being employed as part of the clinical practice. Strategies for neutralizing cytokines by using antibodies, fusion globulins, or soluble receptors have become available and have led to remarkable results in the clinic. Some of these new approaches currently under investigation have led already, or will likely lead shortly, to the registration of new drugs that will supplement existing therapeutic options for inflammatory skin diseases. In addition, the results of the clinical trials are contributing significantly to our understanding of the pathophysiology of diseases and will allow greater insight into which mechanisms play a significant role in their development. Thus, the clinical findings that neutralizing TNF-α is extremely effective in psoriasis, whereas neutralizing IL-8 is not, would suggest a much greater role for TNF-α in the pathogenesis of the disease. Recent research has also led to the discovery of novel cytokines in the expanding IL-1, IL-10, and IL-12 families and has provided insight into the molecular mechanisms of cytokine action, including discovery of novel signal transduction pathways. These investigations are leading to the identification of novel targets that, in contrast to the cytokines and their protein- based neutralizing therapies, may be amenable to inhibition by small molecules that are suitable for oral application. All of these developments may generate additional momentum for still better targeted pharmacological approaches.

To discuss these recent exciting developments we held the workshop: *CYTOKINES AS POTENTIAL THERAPEUTIC TARGETS FOR INFLAMMATORY SKIN DISEASES*, from November 17–19, 2004 in the beautiful environment of the Napa Valley, a short distance north of San Francisco. This meeting was intended to bring together an outstanding group of scientists spanning the broad scope of cytokine research and to have open discussions on recent achievements and future developments in the field. This aim was successfully achieved thanks to the generous support of the Ernst Schering Research Foundation and to the active contributions of the participants. We had the pleasure to host colleagues, who are experts in cytokine genetics, cytokine signaling and skin immunology. At this workshop, sessions comprised: i) Proinflammatory cytokines beyond TNF as novel targets for psoriasis and other immune diseases – lessons from first clinical trials, ii) Modulation of cytokine expression by targeting APC, T-cells and their interactions, iii) Cytokines and their pathways as novel targets for immune diseases and, iv) Inhibition of cytokine signaling mechanisms. The proceedings of the workshop are contained in this volume and we wish to take this opportunity to thank the speakers and participants for their presentations and lively contributions. We trust that the readers will share with us our enthusiasm and continued excitement in the developments coming out of the field of cytokine research, which will open new avenues for the therapy of inflammatory skin diseases and other immune disorders.

Robert Numerof
Charles A. Dinarello
Khusru Asadullah

Contents

1   Inhibition of IFN-γ as a Method of Treatment
of Various Autoimmune Diseases, Including Skin Diseases
B. Skurkovich, S. Skurkovich   1

2   Cytokine Targeting in Psoriasis and Psoriatic Arthritis:
Beyond TNFα
I.B. McInnes   29

3   Inhibitors of Histone Deacetylases
as Anti-inflammatory Drugs
C.A. Dinarello   45

4   Dendritic Cell Interactions and Cytokine Production
M. Foti, F. Granucci, P. Ricciardi-Castagnoli   61

5   Lysosomal Cysteine Proteases and Antigen Presentation
A. Rudensky, C. Beers   81

6   DCs and Cytokines Cooperate for the Induction of Tregs
A.H. Enk   97

7   Induction of Immunity and Inflammation
by Interleukin-12 Family Members
G. Alber, S. Al-Robaiy, M. Kleinschek, J. Knauer,
P. Krumbholz, J. Richter, S. Schoeneberger,
N. Schuetze, S. Schulz, K. Toepfer, R. Voigtlaender,
J. Lehmann, U. Mueller 107

8   The Role of TNFα and IL-17 in the Development
     of Excess IL-1 Signaling-Induced Inflammatory Diseases
     in IL-1 Receptor Antagonist-Deficient Mice
     H. Ishigame, A. Nakajima, S. Saijo, Y. Komiyama, A. Nambu,
     T. Matsuki, S. Nakae, R. Horai, S. Kakuta, Y. Iwakura . . . 129

9   TGFβ-Mediated Immunoregulation
     Y. Peng, L. Gorelik, Y. Laouar, M.O. Li, R.A. Flavell 155

10   TNF Blockade: An Inflammatory Issue
     B.B. Aggarwal, S. Shishodia, Y. Takada,
     D. Jackson-Bernitsas, K.S. Ahn, G. Sethi, H. Ichikawa . . . 161

11   IL-1 Family Members in Inflammatory Skin Disease
     J. Sims, J. Towne, H. Blumberg 187

12   Regulatory T Cells in Psoriasis
     M.H. Kagen, T.S. McCormick, K.D. Cooper 193

Previous Volumes Published in This Series 211

List of Editors and Contributors

Editors

Numerof, R.
Berlex Biosciences,
2600 Hilltop Drive, PO Box 4099, Richmond, CA 94804-0099, USA
(e-mail: Robert_numerof@berlex.com)

Dinarello, C.A.
University of Colorado Health Sciences Center,
4200 East Ninth Ave., B168, Denver, CO 80262, USA
(e-mail: cdinare333@aol.com)

Asadullah, K.
Corporate Research Business Area Dermatology,
Schering AG, Müllerstr. 178, 13342 Berlin, Germany
(e-mail: khusru.asadullah@schering.de)

Contributors

Aggarwal, B.B.
Cytokine Research Laboratory, Department of Experimental Therapeutics,
University of Texas, M.D. Anderson Cancer Hospital,
1515 Holcombe, Box 143, Houston, TX 77030, USA
(e-mail: aggarwal@mdanderson.org)

Ahn, K.S.
Cytokine Research Laboratory, Department of Experimental Therapeutics,
University of Texas, M.D. Anderson Cancer Hospital,
1515 Holcombe, Box 143, Houston, TX 77030, USA

Alber, G.
Institute of Immunology, College of Veterinary Medicine, University of Leipzig,
An den Tierkliniken 11, 04103 Leipzig, Germany
(e-mail: alber@vetmed.uni-leipzig.de)

Al-Robaiy, S.
Institute of Immunology, College of Veterinary Medicine, University of Leipzig,
An den Tierkliniken 11, 04103 Leipzig, Germany

Beers, C.
Department of Immunology, University of Washington School of Medicine,
Box 357370, Health Sciences, Bldg. Rm I-6041 1959 NE Pacific Street, Seattle,
WA 98195, USA

Blumberg, H.
Amgen, Molecular Immunology,
1201 Amgen Court West, Seattle WA 98119, USA

Cooper, K.D.
Case Western Reserve University,
11100 Euclid Avenue Lakeside 3516, Cleveland, OH 44106, USA
(e-mail: kdc@po.cwru.edu)

Enk, A.
Department of Dermatology, University of Heidelberg,
Voßstrasse 2, 69115 Heidelberg, Germany
(e-mail: Alexander.enk@med.uni-heidelberg.de)

Flavell, R.A.
Howard Hughes Medical Institute, Yale University, School of Medicine,
HHMI-310 Cedar, St FMB412, Box 208011, New Haven, CT 06520, USA
(e-mail: Richard.flavell@yale.edu)

Foti, M.
University of Milano-Bicocca, Department of Biotechnology Bioscience,
Piazza della Scienza 2, 20126 Milan, Italy

Gorelik, L.
Biogen Idec, Inc.,
Cambridge, MA 02142, USA

Granucci, F.
University of Milano-Bicocca,Department of Biotechnology Bioscience,
Piazza della Scienza 2, 20126 Milan, Italy

Horai, R.
Center for Experimental Medicine, Institute of Medical Science,
University of Tokyo,
4-6-1 Shirokanedai, Minato-ku, Tokyo 108-8639, Japan

Ishikawa, H.
Cytokine Research Laboratory, Department of Experimental Therapeutics,
University of Texas, M.D. Anderson Cancer Hospital,
1515 Holcombe, Box 143, Houston, TX 77030, USA

Ishigame, H.
Center for Experimental Medicine, Institute of Medical Science,
University of Tokyo,
4-6-1 Shirokanedai, Minato-ku, Tokyo 108-8639, Japan

Iwakura, Y.
University of Tokyo, Center for Experimental Medicine,
Institute of Medical Science,
4-6-1 Shirokanedai, Minato-ku, Tokyo 108-8639, Japan
(e-mail: iwakura@ims.u-tokyo.ac.jp)

Jackson-Bernitsas, D.
Cytokine Research Laboratory, Department of Experimental Therapeutics,
University of Texas, M.D. Anderson Cancer Hospital,
1515 Holcombe, Box 143, Houston, TX 77030, USA

Kakuta, S.
University of Tokyo, Center for Experimental Medicine,
Institute of Medical Science,
4-6-1 Shirokanedai, Minato-ku, Tokyo 108-8639, Japan

Kleinschek, M.
Institute of Immunology, College of Veterinary Medicine, University of Leipzig,
An den Tierkliniken 11, 04103 Leipzig, Germany

Knauer, J.
Institute of Immunology, College of Veterinary Medicine, University of Leipzig,
An den Tierkliniken 11, 04103 Leipzig, Germany

Komiyama, Y.
Center for Experimental Medicine, Institute of Medical Science,
University of Tokyo,
4-6-1 Shirokanedai, Minato-ku, Tokyo 108-8639, Japan

Krumbholz, P.
Institute of Immunology, College of Veterinary Medicine, University of Leipzig,
An den Tierkliniken 11, 04103 Leipzig, Germany

Laouar, Y.
Section of Immunobiology, Yale University School of Medicine,
New Haven, CT 06520, USA

Lehmann, J.
Institute of Immunology, College of Veterinary Medicine, University of Leipzig,
An den Tierkliniken 11, 04103 Leipzig, Germany

Li, M.O.
Section of Immunobiology, Yale University School of Medicine,
New Haven, CT 06520, USA

Mastuki, T.
Center for Experimental Medicine, Institute of Medical Science,
University of Tokyo,
4-6-1 Shirokanedai, Minato-ku, Tokyo 108-8639, Japan

McInnes, I.B.
Division of Immunology, Infection and Inflammation, University of Glasgow,
Centre for Rheumatic Diseases, Glasgow Royal Infirmary,
Alexandra Parade, Glasgow G31 2ER, Scotland
(e-mail: i.b.mcinnes@clinmed.gla.ac.uk)

Mueller, U.
Institute of Immunology, College of Veterinary Medicine, University of Leipzig,
An den Tierkliniken 11, 04103 Leipzig, Germany

Nakae, S.
Center for Experimental Medicine, Institute of Medical Science,
University of Tokyo,
4-6-1 Shirokanedai, Minato-ku, Tokyo 108-8639, Japan

Nakajima, A.
Center for Experimental Medicine, Institute of Medical Science,
University of Tokyo,
4-6-1 Shirokanedai, Minato-ku, Tokyo 108-8639, Japan

Peng, Y.
Department of Rheumatology, University of Washington,
Seattle, WA, 98106, USA

Ricciardi-Castagnoli,P.
Department of Biotechnology Bioscience, University of Milano-Bicocca,
Piazza della Scienza 2, 20126 Milan, Italy
(e-mail: paoloa.castagnoli@unimib.it)

Richter, J.
Institute of Immunology, College of Veterinary Medicine, University of Leipzig,
An den Tierkliniken 11, 04103 Leipzig, Germany

Rudensky, A.
Howard Hughes Medical Institute, University of Washington
School of Medicine,
Box 357370, Health Sciences, Bldg. Rm I-6041 1959 NE Pacific Street, Seattle,
WA 98195, USA
(e-mail: aruden@u.washington.edu)

Saijo, S.
University of Tokyo, Center for Experimental Medicine,
Institute of Medical Science,
4-6-1 Shirokanedai, Minato-ku, Tokyo 108-8639, Japan

Schoeneberger, S.
Institute of Immunology, College of Veterinary Medicine, University of Leipzig,
An den Tierkliniken 11, 04103 Leipzig, Germany

Schuetze, N.
Institute of Immunology, College of Veterinary Medicine, University of Leipzig,
An den Tierkliniken 11, 04103 Leipzig, Germany

Schulz, S.
Institute of Immunology, College of Veterinary Medicine, University of Leipzig,
An den Tierkliniken 11, 04103 Leipzig, Germany

Sethi, G.
Cytokine Research Laboratory, Department of Experimental Therapeutics,
The University of Texas, M.D. Anderson Cancer Center,
Box 143, 1515 Holcombe Boulevard, Houston, Texas 77030, USA

Shishodia, S.
Cytokine Research Laboratory, Department of Experimental Therapeutics,
The University of Texas, M.D. Anderson Cancer Center,
Box 143, 1515 Holcombe Boulevard, Houston, Texas 77030, USA

Sims, J.E.
Amgen, Molecular Immunology,
1201 Amgen Court West, Seattle WA 98119, USA
(e-mail: sims@amgen.com)

Skurkovich, B.
Pediatric Infectious Diseases, Rhode Island Hospital,
593 Eddy Street, Providence, RI 2903, USA
(e-mail: Bskurkovich@pol.net)

Skurkovich, S.
Advanced Biotherapy, Inc.
802 Rollins Ave. Rockville, MD 20852, USA
(e-mail: sskurkovich@erols.com)

Straubinger, R.K.
Institute of Immunology, College of Veterinary Medicine, University of Leipzig,
An den Tierkliniken 11, 04103 Leipzig, Germany

Takada, Y.
Cytokine Research Laboratory, Department of Experimental Therapeutics,
The University of Texas M.D. Anderson Cancer Center,
Box 143, 1515 Holcombe Boulevard, Houston, Texas 77030, USA

Toepfer, K.
Institute of Immunology, College of Veterinary Medicine, University of Leipzig,
An den Tierkliniken 11, 04103 Leipzig, Germany

Towne, J.
Amgen, Molecular Immunology,
1201 Amgen Court West, Seattle WA 98119, USA

Voigtlaender, R.
Institute of Immunology, College of Veterinary Medicine, University of Leipzig,
An den Tierkliniken 11, 04103 Leipzig, Germany

1 Inhibition of IFN-γ as a Method of Treatment of Various Autoimmune Diseases, Including Skin Diseases

B. Skurkovich, S. Skurkovich

1.1	Introduction	2
1.2	Treatment of Th-1-Mediated Autoimmune Diseases	4
1.2.1	Rheumatoid Arthritis	5
1.2.2	Multiple Sclerosis	5
1.2.3	Corneal Transplant Rejection	6
1.2.4	Type I Diabetes	6
1.2.5	Uveitis	7
1.3	Treatment of Th-1-Mediated Autoimmune Skin Diseases Using Anti-IFN-γ	7
1.3.1	Psoriasis Vulgaris	7
1.3.2	Alopecia Areata	9
1.3.3	Vitiligo	10
1.3.4	Acne Vulgaris	11
1.3.5	Herpes Simplex Virus Type 1	11
1.4	Genetic Skin Diseases in Which Cytokines May Be Involved	12
1.4.1	Dystrophic Epidermolysis Bullosa (DEB)	12
1.5	Other Th-1-Mediated Skin Diseases in Which Testing of Anticytokine Therapy Is Warranted	13
1.6	Beneficial Effect of Anti-IFN-γ in Possibly Non-Th-1-Mediated Skin Diseases	14
1.6.1	Herpes Simplex Virus Type 2	14

1.7　　Skin Diseases Involving Th-1/Th-2 Cytokines 14
1.8　　Discussion 15
References 19

Abstract. We pioneered anticytokine therapy (ACT) for autoimmune diseases (ADs). In 1974, we proposed that hyperproduced interferon (IFN) can bring AD and anti-IFN can be therapeutic. In 1989, we proposed removing tumor necrosis factor (TNF)-α together with certain types of IFN to treat various ADs. We found IFN in patients with different ADs and conducted the first clinical trial of ACT in 1975. Anti-IFN-γ and anti-TNF-α work in similar ways, but the latter brings serious complications in some patients. We obtained good, sometimes striking, therapeutic effects treating many different Th-1-mediated ADs with anti-IFN-γ, including rheumatoid arthritis, multiple sclerosis (MS), corneal transplant rejection, and various autoimmune skin diseases such as psoriasis, alopecia areata, vitiligo, acne vulgaris, and others. Anti-IFN-γ was in some ways superior to anti-TNF-α, which was ineffective in MS. Anti-IFN-γ therapy holds great promise for treating many Th-1 ADs, especially skin diseases.

1.1　Introduction

Anticytokine therapy was pioneered by S. Skurkovich, who in 1974 published in *Nature* the proposal that the hyperproduction of interferon (IFN) (a cytokine) can bring autoimmune disease and that neutralization of IFN could be therapeutic (Skurkovich et al. 1974). This proposal was based on data describing the mechanism of IFN action (Skurkovich et al.1973). It was postulated that IFN production was a part of the immune response and that any disturbance of IFN production could lead to immune system dysregulation and vice versa (Skurkovich et al. 1973, 1982). In other words, a disturbance of IFN synthesis can bring disease. In 1975, IFN was found in the blood of autoimmune patients and the first anticytokine therapy was tested (Skurkovich and Eremkina 1975).

Many years ago, it was proposed that anticytokine therapy could be beneficial in treating a wide variety of autoimmune diseases and diseases of supposed autoimmune genesis, in which disturbed IFN synthesis is a common mechanism of pathology; these included prolongation of skin allografts, collagenoses, rheumatism, hemolytic anemia, immune form of idiopathic thrombocytopenic purpura, eczemas, nephritis, myasthenia gravis, Hashimoto disease, autoimmune diseases of the organs of sight and hearing, multiple sclerosis, Menière's disease, Parkinson's disease, pemphigus, schizophrenia and possibly some mental derangements, psoriasis and Crohn's disease (Skurkovich et al. 1974, 1975, 1977, 1991). This was considered a radical proposal at the time, but since then the literature supporting use of anticytokine therapy as a universal treatment in various Th-1-mediated autoimmune diseases has accumulated.

Studies in experimental rodent models of autoimmune diseases confirm the hypothesis of the role of IFN-γ in autoimmunity. In these models, IFN-γ can trigger or exacerbate disease, as in thyroiditis, anterior uveitis, and autoimmune neuritis (Kawakami et al. 1990; Egwuagu et al. 1999; Hartung et al. 1990), and anti-IFN-γ can bring beneficial results, such as in presenting type I diabetes (Debray-Sachs et al. 1991; Nicoletti et al. 1997), and treating experimental autoimmune thyroiditis (Tang et al. 1993),and neuritis (Tsai et al. 1991). The administration of IFN-α and -γ to patients for the treatment of cancer and other diseases has led to the triggering or exacerbation of a great variety of underlying autoimmune diseases (Miossec 1997). It was also found that mice deficient in the IFN-γ receptor are less susceptible to collagen-induced arthritis (Kageyama et al. 1998). Thus, in the laboratory, research supports our hypothesis that disturbed IFN-γ production can lead to disease. The central role of IFN-γ in the pathogenesis of many organ-specific autoimmune diseases will be discussed below.

In 1989, it was clear that IFN-γ induces other cytokines, such as TNF-α. We first proposed to remove TNF-α together with pH labile IFN-α in autoimmune diseases and AIDS, which is also a disease with autoimmune components (Skurkovich et al. 1989). (IFN-γ is pH labile, while IFN-α is normally pH stable.) Now TNF-α antagonists have been commercialized by several companies to treat several autoimmune diseases, including rheumatoid arthritis, Crohn's disease, and psoriasis.

Though we were the first to propose treating autoimmune patients with anti-TNF-α, we have particularly investigated use of anti-IFN-γ; at the same time we have compared the effectiveness of anti-IFN-γ and anti-TNF-α in various autoimmune diseases. Our investigations have shown that anti-IFN-γ may have clinical effects that are as good as and sometimes superior to anti-TNF-α. Based on the differentiation of T helper cells into Th-1, Th-2, Th3, and Th0 subsets (Mosmann and Sad 1996), autoimmune and allergic diseases are sometimes distinguished as Th-1 or Th-2 diseases, respectively, depending on whether they are mediated by a polarization of T helper cells toward the Th-1 or Th-2 subset. Organ-specific autoimmune diseases, considered here, are generally characterized by the production of Th-1, or proinflammatory cytokines, including IFN-γ, IL-2, TNF-β, and others. TNF-α is also considered a proinflammatory cytokine.IFN-γ and TNF-α in many cases work synergistically and have many of the same effects.

Based on our hypothesis and accumulating data in the laboratory and clinic supporting the importance of IFN-γ and TNF-α in the pathogenesis of Th-1-mediated conditions, we have clinically tested the use of anticytokine therapy, in this case anti-IFN-γ and sometimes anti-TNF-α antibodies, in the treatment of a broad range of these diseases. We present here our data with particular attention to skin diseases. We also will review the laboratory evidence supporting this approach.

1.2 Treatment of Th-1-Mediated Autoimmune Diseases

We conducted the first anti-IFN-γ therapy in Th-1-mediated diseases together with several clinics using a polyclonal anti-IFN-γ antibody (both IgG and F(ab')2 antibody fragments) and in some cases, for comparison, a polyclonal anti-TNF-α. The antibodies were given intramuscularly once or twice daily for 5–6 days. Our studies include placebo-controlled, double-blind studies in patients with rheumatoid arthritis unresponsive to standard treatments (Sigidin et al. 2001) and in patients with secondary progressive multiple sclerosis (Skurkovich et al. 2001). In addition, we conducted smaller proof-of-principle studies in patients experiencing corneal transplant rejection (Skurkovich et al. 2002a), in psoriatic arthritis (Skurkovich et al. 1998), type I diabetes (Skurkovich et al. 2003a), uveitis (Skurkovich et al. 2003b), juvenile rheumatoid arthritis, aukylosing spondylitis and in several autoimmune skin diseases (Skurkovich et al. 2002b).

1.2.1 Rheumatoid Arthritis

An open-label study (23 patients) (Skurkovich et al. 1998) and subsequent double-blind, placebo-controlled studies (55 patients) of the use of anti-IFN-γ compared to placebo and/or anti-TNF-α in patients with RA demonstrated statistically significant improvement in the key symptoms of the disease, using American College of Rheumatology (ACR) criteria (Skurkovich et al. 1998; Sigidin et al. 2001; Lukina et al. 2003; Lukina 2004). In the controlled study, anti-IFN-γ was compared to a placebo as well as to anti-TNF-α in patients followed for 28 days. Results demonstrated that improvement with the use of anti-IFN-γ was as good as and sometimes superior to anti-TNF-α with fewer side effects and higher remission rates (Lukina 2004). Ultrasound data indicated significant reduction of swelling in the synovial joint membrane in the anti-IFN-γ group only (Lukina et al. 2003).

Commercially available TNF-α antagonists have brought many serious complications, such as systemic lupus erythematosus, reactivation of latent TB, liver problems, alopecia, lymphoma, and others (Michel et al. 2003; Cunnane et al. 2003; CDC 2004; Ettefagh et al. 2004; Hamilton 2004).

Giving IFN-γ to patients with rheumatoid arthritis has been tried several times with varied results, including a positive effect, no effect, and exacerbation of the arthritis in comparison to a placebo (Cannon et al. 1993; Veys et al. 1997; Seitz et al. 1988). Given our positive results with an anti-IFN-γ antibody, we speculate that positive results may be due to a suppressive effect on IFN-γ production by the administration of IFN-γ.

1.2.2 Multiple Sclerosis

Anti-IFN-γ and anti-TNF-α antibodies were also used in an open-label study (22 patients), followed by a double-blind, placebo-controlled (45 patients) trial in patients with secondary progressive multiple sclerosis in which patients were followed for 1 year (Skurkovich et al. 2001). In the controlled trial, the group receiving anti-IFN-γ showed a significant increase in the number of patients free of sustained EDSS (Expanded Disability Status Scale) progression and in mean time without sustained

EDSS progression as well as a significant increase in the number of patients without gadolinium-enhancing lesions using magnetic resonance imaging (MRI). No significant improvement was found using anti-TNF-α, which confirms other studies in multiple sclerosis patients in which anti-TNF-α was found ineffective (Lenercept Multiple Sclerosis Study Group 1999).

1.2.3 Corneal Transplant Rejection

Evidence supports the idea that reducing IFN-γ could have a beneficial effect on corneal graft survival. For example, mice with immune responses manipulated to respond preferentially toward a Th-2 pathway experienced a sharp drop in the production of the Th-1 cytokine IFN-γ, and a greater than 50% reduction in corneal graft rejection (Yamada et al. 1999).

In our study of patients suffering corneal transplant rejection after penetrating keratoplasty, 2–3 days after the start of a 6-day treatment course in which eye drops were administered 3–4 days a time, transplant transparency improved, and edema decreased in 10 of 13 patients (Skurkovich et al. 2002a). At the end of week 1, the transplant became almost fully transparent, eye inflammation disappeared, and visual acuity improved. In the remaining three patients, additional treatment courses were needed for comparable improvement. The improvement was maintained through follow-up averaging of 7 months. Four patients with corneal transplant rejection given anti-TNF-α also had comparable positive results (Kasparov et al. 2004). We plan to test our approach in transplant rejection of various organs using a monoclonal anti-IFN-γ or anti-TNF-α, soluble receptors to IFN-γ or TNF-α or anti-CD20.

1.2.4 Type I Diabetes

In a small group of three type I diabetic patients, a single 6-day course of anti-IFN-γ antibodies was associated with a rise in C-peptide levels, a drop in blood glucose, and a reduction in the insulin dose (Skurkovich et al. 2003a). Larger clinical studies are planned using a humanized monoclonal antibody to IFN-γ, to TNF-α, and possibly both anti-IFN-γ and anti-TNF-α together.

1.2.5   Uveitis

In six patients with juvenile rheumatoid arthritis-associated uveitis, anti-IFN-γ was found to be effective as an adjunct therapy in a comparison of standard therapy plus anti-IFN-γ vs standard therapy alone (Skurkovich et al. 2003b). In four of six patients, anti-IFN-γ when used with standard therapy appeared to halve the duration of the acute period, decrease the severity of symptoms during the acute period, and substantially increase the length of remission in these patients compared to standard therapy alone.

1.3   Treatment of Th-1-Mediated Autoimmune Skin Diseases Using Anti-IFN-γ

We used anti-IFN-γ in patients with psoriasis vulgaris, alopecia areata, vitiligo, acne vulgaris, and herpes simplex type 1, and plan to expand the trials using a humanized monoclonal antibody. Trials are also planned for testing the treatment in other Th-1-mediated skin diseases, including seborrheic dermatitis, rosacea, oral lichen planus, and dermatophytosis, which are discussed in the following sections. In most cases, we used two methods of delivery of the antibody: intradermal injection around the lesion and intramuscular administration. Thus, anti-IFN-γ delivered to the lesion either IM or locally brought the same effect. We believe anticytokine therapy, particularly the removal or inhibition of IFN-γ and/or TNF-α and IL-1, may be a rational treatment for many severe autoimmune skin diseases that do not respond to standard therapies.

1.3.1   Psoriasis Vulgaris

We first proposed that anticytokine therapy could be beneficial in treating psoriasis in 1977 (Skurkovich et al. 1977). Recently in an open-label clinical trial using anti-IFN-γ in patients with psoriasis vulgaris using both intramuscular and intradermal injections, we observed very rapid, sometimes striking responses. Seven patients who received IM injections of anti-IFN-γ experienced a very rapid reduction of the intensity of the erythema, leading to the disappearance of papular infiltrates and, after a few weeks, clearing of the psoriatic plaques and complete remission in four of the patients (see Fig. 1). Three who still had small infiltrates on the legs were given UV treatments (200–500 nm wavelength), after which two went into full remission and one completed therapy with 80% reduction in plaques (Skurkovich et al. 2002b, 2002c, 2004a). In addition, three patients were given intradermal injections of the antibody fragments (1 ml aliquoted into ten 0.1-ml injections) daily for 7 days around the site and in the center of one lesion. The improvement in these lesions followed a similar course as those of the first seven patients with improvement in nearby untreated lesions noted in some cases. Psoriasis also improved in one patient with psoriasis in addition to type I diabetes when treated with anti-IFN-γ for the diabetes. TNF-α antagonists are already used commercially for the treatment of psoriasis but require caution because of possible infectious complications.

Psoriasis is now generally considered to be a Th-1-mediated autoimmune disease, with evidence implicating the proinflammatory cytokines IFN-γ and TNF-α (Ozawa and Aiba 2004). IFN-γ has also been shown to trigger psoriasis (Fierlbeck et al. 1990), and psoriatic epidermal T cells produce and secrete IFN-γ within the lesion (Koga et al. 2002). The level of IFN-γ in the sera of psoriasis patients has been found to positively correlate with all indices of disease severity (Jacob et al. 2003).

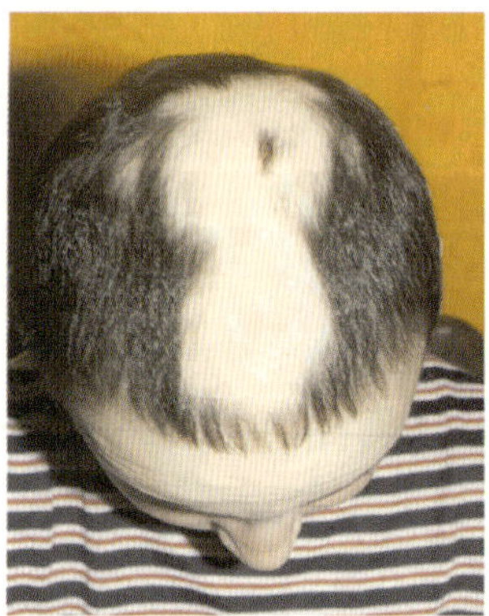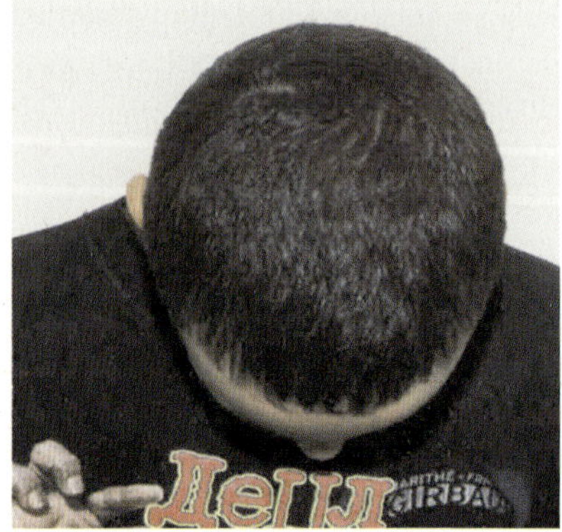

Fig. 1. Eleven-year-old patient before treatment with anti-interferon-gamma antibodies (*left*). Patient 14 days after start of treatment (*right*)

1.3.2 Alopecia Areata

T-cell clones from alopecia lesions have been reported to release large amounts of IFN-γ and/or TNF-α that downregulate epithelial cell proliferation (Thein et al. 1997).

We obtained encouraging results using anti-IFN-γ in treating 16 patients (age 6–15 years) with AA in the progressive stage (Skurkovich 2002b, 2002c, 2003c, 2004b; Sharova 2003). Nine had patchy, progressive disease (based on a positive hair pull test at the periphery of the patches), five patients had 100% scalp hair loss, some with eyebrows and/or eyelashes preserved, and two patients had patchy, stable hair loss. Of the 16 patients treated, eight of the nine patients with patchy, progressive hair loss stabilized after 3 days of therapy and showed no additional hair loss. After 4–6 months, these patients showed partial, but on-going (four patients) or full (four patients) terminal, i.e., fully restored, hair-growth, including eyebrows and/or eyelashes, with one having no response (see Fig. 2). By 6 months, only one of these patients had any recurrence (one 1.5-cm lesion that appeared 5 months after

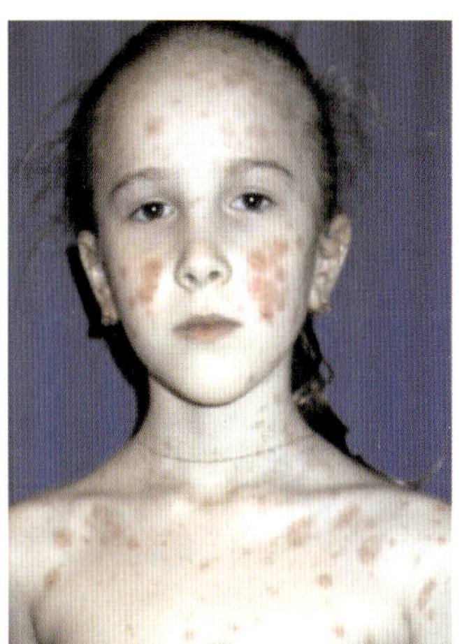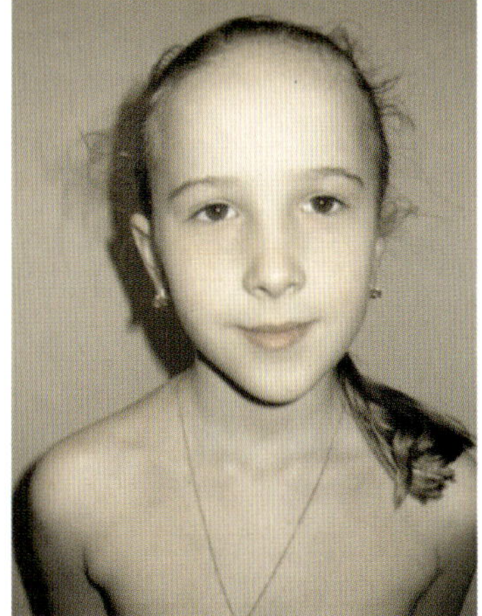

Fig. 2. *Left:* Fourteen-year-old patient with alopecia areata lasting 4.5 months before intradermal treatment with anti-IFN-γ antibodies. *Right:* Patient 12 weeks after start of treatment

treatment), with the rest maintaining or continuing their improvement. Of the five patients with 100% baldness, three had some limited terminal hair growth; the other two showed only vellus hair or no response. Of the two patients with stable but patchy hair loss, one showed complete eyelash and eyebrow hair restoration but both showed no hair growth in the scalp. The eight patients with progressive hair loss were followed for 2 years with two recent relapses. The results support early intervention as most effective in preventing progression to total hair loss. The immediacy and high rate of response to the injections in progressive cases argue against spontaneous remission of the disease as an explanation for the improvement, but larger, controlled trials are planned using a humanized monoclonal antibody.

Since our trial, more experimental evidence has strongly implicated IFN-γ in the pathogenesis of AA. Researchers presenting results at the Alopecia Workshop of the 4th Intercontinental Meeting of Hair Research Societies in Berlin, induced AA in wild-type and IFN-γ knockout mice by grafting lesional AA skin from alopecia-affected mice. The knockout mice were found to be resistant to the development of AA compared to 90% of the wild-type mice, who developed alopecia. It was hypothesized that, with no IFN-γ produced, T-cells were not activated in the knockout mice, confirming that IFN-γ may play a key role in activating autoreactive T-cells (Freyschmidt-Paul 2004). Alopecia areata (AA) is now discussed as an autoimmune disease of the hair follicle, especially expressing the Th-1 cytokines interleukin-1β, interleukin-2, and interferon-γ (Arca et al. 2004). Progressive AA patients' peripheral blood mononuclear cells display an increased percentage of activated T cells and IL-12 and IFN-γ expression (Zoller et al. 2004), a phenomenon frequently observed in autoimmune diseases (Chinon and Shearer 2003).

1.3.3 Vitiligo

Of four vitiligo patients who received intradermal injections around the lesion, two responded with repigmentation in the treated area (Skurkovich et al. 2002b, 2002c, 2003c). In the other two patients, an additional course of intramuscular injections led to a gradual diminishment of the border between the depigmented area and normal skin. Vitiligo lesions also cleared up in a patient treated for AA with anti-IFN-γ. As with psoriasis and alopecia, vitiligo is considered a Th-1-mediated autoimmune disease. Perilesional T-cell clones derived from patients with vitiligo show a predominant Th-1-like cytokine secretion profile (Wankowicz-Kalinska et al. 2003), and vitiligo patients demonstrate a statistically significant increase in the expression of IFN-γ in involved and adjacent uninvolved skin (Grimes et al. 2004).

1.3.4 Acne Vulgaris

IFN-γ may play a central role in the immunopathogenesis of acne (Mouser et al. 2003). The normal skin commensal bacterium *Propionibacterium acnes* (P. acnes), which is implicated in the pathogenesis of inflammatory acne, appears to stimulate a typical Th-1 response. T cell lines from inflamed acne lesions were shown to proliferate in response to *P. acnes* extract and expressed IFN-γ in response to *P. acnes* stimulation (Mouser et al. 2003).

In a small study, patients with acne vulgaris with multiple infiltrated pustules and elements on the cheeks or forehead were treated with topical anti-IFN-γ. A wet gauze compress soaked with anti-IFN-γ in liquid form was applied with mild pressure to the affected areas for 1 min, three times a day, for 4 days. By the 2nd day, nearly all pustular elements had dried up with no fresh elements appearing. After 4 days of treatment, the infiltrated elements remained but had paled in color. In another patient, fresh pustules in smaller quantity than before appeared on the 5th day after treatment. Further studies are warranted, and in future, for severe acne, we plan to prepare a complex of anti-IFN-γ, an antibiotic, and a retinoid and to test different methods of delivery, including oral, topical, and parenteral.

1.3.5 Herpes Simplex Virus Type 1

Herpes simplex is a common virus affecting the skin, mucous membranes, nervous system, and the eye. Herpes simplex type 1 virus has been found to induce certain autoimmune conditions. Autoreactive T cells triggered by herpes simplex type 1 virus play an important role in the pathogenesis of herpes-associated erythema multiforme (HAEM) (Aurelian et al. 2003), in which IFN-γ is expressed in the lesions (Kokuba et al. 1999). Herpes stromal keratitis, an autoimmune disease characterized by T-cell-dependent destruction of the corneal tissue, is thought to be due to an epitope expressed by a coat protein of the herpes simplex virus being recognized by autoreactive T cells that target corneal antigens in a mouse model (Zhao et al. 1998). Herpes simplex virus type 1 lesions of the mouth and lips may also be Th-1-mediated. In preliminary studies, we found a close association between topical application of anti-IFN-γ to herpes lesions and reduction in burning and pain in the lesion area. Thus removal of IFN-γ or possibly TNF-α or IL-1 could be beneficial.

Patients with severe, chronic recurring herpes simplex on the lips presenting with erythema and blisters unresponsive to standard treatment were treated with an anti-IFN-γ-soaked gauze compress applied to the lesions every 2 h for 1–2 min. After 3–3.5 h, the patients noted rapid reduction of the burning and pain in the area of the lesion. After 2 days, scabs formed, which quickly faded in 5 days. In future, different methods of delivery, including topical (especially on the eyes), oral, and parenteral will be tested.

1.4 Genetic Skin Diseases in Which Cytokines May Be Involved

1.4.1 Dystrophic Epidermolysis Bullosa (DEB)

DEB is a genetic disease involving a defect in the gene for type VII collagen (Gardella et al. 1996). However, it may also involve autoimmune mechanisms in the period of mutation or afterward. IFN-γ could also be involved by altering gene expression (Buntinx et al. 2004). Based on evidence in the literature of immune system involvement and signs of inflammation in a patient suffering severely from DEB, we treated two cases with striking results. For example, in one case a 14-year-old boy with DEB from birth was given a course of anti-IFN-γ antibodies intramuscularly. Before treatment, his temperature was 39.2°C. Skin on the back of his neck, lower back, and upper and lower extremities was covered with erosive, ulcerative lesions up to 15 cm in diameter with pustular, hemorrhagic scabs. Around the lesions were painful areas of hyperemia and swelling, on the mucous membranes of the mouth and genitals were erosive lesions, and nails were absent from the fingers and toes. After the second injection of anti-IFN-γ, his temperature normalized. On day 2 after the start of treatment, the pain, swelling, and hyperemia around the ulcerative lesions at the back of the neck disappeared, as did signs of infectious damage to the skin on the back. By day 5 of the treatment, the erosive lesions on the mucous membrane of the mouth epithelialized. Seven days after the start of treatment, active epithelialization of the ulcerative skin lesions was observed. The child began to eat, his general sense of well-being significantly improved, and his sleep normalized (Korotky et al. 2004).

Very low levels of IFN-γ, IL-1, and IL-2 have been found to be produced in vitro by peripheral blood mononuclear cells from patients with severe forms of EB (recessive dystrophic and dominant dystrophic) as compared to sex- and age-matched controls (Chopra et al. 1992). Low cytokine production in vitro is typical of autoimmune diseases and may in those cases be due to exhaustion from overproduction of these cytokines. In addition, persons with more severe forms of EB have demonstrated significant reductions in natural-killer (NK) cell activity (Tyring et al. 1989). Deficiencies in NK cells have been shown to be associated with certain autoimmune diseases (Baxter and Smyth 2002). This case brought us strong emotional satisfaction. In future, we plan to test the treatment using humanized monoclonal antibodies to various cytokines and anti-CD20.

1.5   Other Th-1-Mediated Skin Diseases in Which Testing of Anticytokine Therapy Is Warranted

Other skin diseases that appear to involve a Th-1-mediated immune response include seborrheic dermatitis (SD), rosacea, oral lichen planus, and dermatophytosis. SD has upregulated expression of proinflammatory cytokines in skin biopsies of SD lesions (Molinero et al. 2003). T-cell subsets are also found in the dermal infiltrates of rosacea lesions (Rufli and Buchner 1984), and IFN-γ-producing cells are found in dermatophytosis (Koga et al. 2001). In oral lichen planus, mononuclear cells express IFN-γ in the superficial lamina propria (Khan et al. 2003). Our near-term goal is to test anticytokine therapy in these conditions.

1.6 Beneficial Effect of Anti-IFN-γ in Possibly Non-Th-1-Mediated Skin Diseases

1.6.1 Herpes Simplex Virus Type 2

IFN is generally known to prevent viral replication. Recurrent herpes simplex virus type 2 genital lesions, on the other hand, may be a special case in which dysregulated production of IFN-γ appears to exert a pathological effect on the organism by paradoxically helping the virus survive or replicate (Singh et al. 2003). We have proposed that some viruses may belong to a special category in which IFNs or TNF-α production is dysregulated, and in some cases these cytokines may help sustain the virus (Skurkovich et al. 1987). In this category may belong oncogenic viruses, the HIV, rubella virus, papillomavirus, possibly certain other viruses and bacteria, and some chemical agents. Herpes simplex virus type 2 may also belong in this class. IFN-γ may also directly damage cells in this disease. Thus, removal of IFN-γ, TNF-α, or IL-1 may help destroy the viruses or other agents sustained by these cytokines and ameliorate the condition. We have very preliminary positive data from the topical application of anti-IFN-γ in a patient with a herpes simplex virus type 2 lesion. Pain was immediately relieved upon application of the antibody, and the skin began to quickly epithelialize. Further studies using anti-IFN-γ need to be undertaken to confirm this very preliminary finding and in addition to test topical use of antibodies to TNF-α and IL-1 in different combinations in this condition. Parenteral and oral administration of a special form of the preparation should also be tested.

1.7 Skin Diseases Involving Th-1/Th-2 Cytokines

Atopic dermatitis is a disease in which Th-1 and Th-2 cytokines are involved depending on the phase of the disease. IFN-γ-positive cells are predominant in the skin lesions. In the initiation phase, IL-4 production by Th-2 and Th-0 cells is predominant over IFN-γ production by Th-1 cells, while in the late and chronic phase, the situation is reversed with IFN-γ production dominant over IL-4, IL-5, and IL-13 production by Th-2 and Th-0 cells (Thepen et al. 1996). Thus, in the initiation phase, antibodies to IL-4, IL-5, and IL-13 alone or in combination could be beneficial while use of anti-IFN-γ may be rational in the later phase of this disease. According to the literature, anti-TNF-α therapy has, however, triggered atopic dermatitis (Mangge et al. 2003). In this case, the anti-TNF-α may have been administered during the wrong phase for such treatment. Blocking Th-2 cytokines may be beneficial in the initiation phase.

1.8 Discussion

IFN-γ is a cytokine released mainly by certain T cells and natural killer cells. It drives the immune response by directing CD4$^+$ T cells toward a Th-1 phenotype, activating macrophages to kill pathogens, enhancing or inducing MHC class I and class II molecules and stimulating B cells to mature and secret antibodies (Snapper 1996). In many autoimmune diseases, IFN-γ appears to play a key role in activating autoreactive T cells (Buntinx et al. 2002). Thus different clinical manifestations of these diseases may depend on the cell territory in which these cytokines are hyperproduced (Skurkovich et al. 1994). High levels of IFN-γ or IFN-γ-producing T cells compared to controls have been found in the disease sites of many Th-1 autoimmune diseases, such as in the synovial fluid in rheumatoid arthritis (Canete et al. 2000), in the cerebrospinal fluid and plaques in multiple sclerosis (Woodroofe and Cuzner 1993; Traugott and Lebon 1988), in the aqueous humor and corneal infiltrating cells in cases of corneal transplant rejection (Yamagami et al. 1998), in the islet β cells in type I diabetes (von Herrath and Oldstone 1997), in the ocular cells in uveitis (Whitcup et al. 1992), and in the lesions of various autoimmune skin diseases, such as psoriasis (Barker et al. 1991), vitiligo (Grimes et al. 2004), and acne (Mouser et al. 2003). Though the exact mechanism of autoimmunity is not well understood, evidence is rapidly accumulating that IFN-γ, IFN-α, and TNF-α may not only participate in the disease process but also directly increase disease progression. As mentioned, we first found circulating IFNs in patients with various autoimmune diseases (Skurkovich et al. 1975). In humans, IFN-γ can, for example, trigger or exacerbate multiple sclerosis (Panitch et al. 1987), rheumatoid arthritis (Seitz et al. 1988), psoriasis (Fierlbeck et al. 1990), and others. IFN-α can also induce underlying autoimmune diseases, including autoimmune thyroid disease (Murakami et al. 1999), type I diabetes (Fabris et al. 1998), and rheumatoid arthritis (Pittau et al. 1997). In multiple sclerosis, evidence points to IFN-γ as having a direct role in inducing central nervous system demyelination (Horwitz et al. 1997). IFN-γ in chronic high levels may set in motion the autoimmune process. Some pathological action is mediated by the induction by IFN-γ of nitric oxide synthetase, which produces nitric oxide (NO), which can cause neurodegeneration (Moncada et al. 1991). NO is produced in large amounts by activated macrophages in response to high levels of IFN-γ acting synergistically with other cytokines such as TNF-α and IL-1, in whose induction IFN-γ participates (Goodwin et al. 1995). NO damages melanocytes in vitiligo (Rocha and Guillo 2001) and β cells in diabetes (Thomas et al. 2002).

Certain endogenous and exogenous factors (genetic, viruses, bacteria, fungi, in some cases sex hormones, parasites, and others) may be involved in the initiation of autoimmune diseases and may induce IFN-γ. Scientists have implicated Epstein-Barr virus, mycobacteria, Proteus, and *Escheria coli*, for example, in the pathogenesis of rheumatoid arthritis (Harrison and McColl 1998). Human endogenous retroviruses (HERV) have been found in patients with multiple sclerosis (Christensen et al. 2001) and schizophrenia (Karlsson et al. 2001). Coxsackie viruses may help bring about type I diabetes (Harrison and McColl 1998). Certain pediatric autoimmune neuropsychiatric diseases are associated with streptococcus (PANDAS), including obsessive-compulsive disorder and Tourette's syndrome (Snider and Swedo 2003). A protein of Group A β-hemolytic streptococci has been implicated in psoriasis (Prinz 1997). Certain other skin diseases also appear to be associated with particular viruses, bacteria, fungi, possibly parasites, and sex hormones that may induce Th-1 cytokines. As mentioned, IFN-γ, TNF-α, and Th-1-positive cells or mRNA expression of IFN-γ and other Th-1 cytokines are found in skin lesions of autoimmune skin diseases, such as alopecia, psoriasis, and vitiligo. *P. acnes*, implicated in acne, when injected into horses, increases IFN-γ expression (Davis et al. 2003). Sex hormones also act in acne, possibly as inducers of IFN-γ. Since estrogen upregulates IFN-γ production (Karpuzoglu-Sahin et al. 2001), hormones may act together with *P. acnes* to increase IFN-γ production. It is possible to consider acne a temporary autoimmune condition that may depend in part on the effects of high levels of sex hormones in such periods as adolescence, the premenstural period, and pregnancy. SD is associated with the fungus *Pityrosporum ovale* (Malassezia furfur) (Faergemann et al. 1996). IFN-γ has been implicated in SD through the detection of IFN-γ induced protein 10 (IP-10) in keratinocytes in biopsies of SD lesions (Smoller et al. 1990). Antimycotics or antibiotics are effective to some extent in SD (Reichrath 2004), possibly because by removing the fungus, one removes the possible inducer of proinflammatory cytokine production. IFN-γ-positive cells are also present in the skin lesions of dermatophytosis (Koga et al. 2001), and in rosacea T-cell subsets in the dermal infiltrates of lesions are frequently associated with the parasite *Demodex folliculorum* (Rufli and Buchner 1984). Though there is little research on this parasite as an inducer of proinflammatory cytokines, chronic dermatitis was found associated with large numbers of Demodex ectoparasites, and pronounced CD4 and CD8 T-cell infiltrates in the dermis and lymphadenopathy associated with increased IFN-γ and IL-12 expression were observed (Liu et al. 2004).

Drugs that have some positive effect in certain skin diseases inhibit proinflammatory cytokines IFN-γ and/or TNF-α, which may be behind the effective action of these drugs. Acne and psoriasis respond to vitamin A, which inhibits IFN-γ release by peripheral blood mononuclear cells (Wauben-Penris et al. 1998). The drug pimecrolimus, used to treat atopic dermatitis and psoriasis, also inhibits synthesis of IFN-γ and TNF-α (Wolff and Stuetz 2004). Topical tacrolimus, effective in bringing about repigmentation in vitiligo, appears to work by suppressing TNF-α production (Grimes et al. 2004). In a psoriasis mouse model, mechlorethamine, a drug that inhibits IFN-γ, TNF-α/β, and IL-12, inhibited psoriasis (Tang et al. 2003). Certain natural factors may help to stabilize IFN-γ production, such as vitamins A and D (Cippitelli and Santoni 1998; Muller and Bendtzen 1996). This action of vitamin D may be an important factor in multiple sclerosis, in which a gradient of increasing prevalence with increasing latitude has been observed; the lower prevalence in southern climates is attributed to an ultra-violet radiation-induced increase in serum vitamin D levels, which may have a protective effect (Ponsonby et al. 2002). Normal intake of vitamins, minerals and microelements from food possibly helps maintain the balance between Th-1 and Th-2 cytokine production – possibly one of the functions of vitamins in the human and other living organisms (Albers et al. 2003). This suggests specific targeting of IFN-γ or TNF-α, and IL-1 may have an even stronger, positive effect in treating these diseases.

Thus, IFN-γ plays a fundamental role in the regulation of the metabolism and immunological status of cells in a healthy state. A disturbed IFN-γ production with hyperproduced IFN-γ is found in different sites in the organism and can bring many somatic and neuropsychiatric diseases. In healthy people, IFN-γ is not found in the blood. The disturbance of IFN-α production in addition to other pathological actions may, as we have hypothesized before (Skurkovich et al. 1993, 1994, 1998, 2002d), also play a key role in the development of the autoantigen and the autoantigen complex. We have proposed that possibly, besides IFN-α, IFN-γ may play a similar role (Skurkovich et al. 2002d). There may exist more than one type of IFN connected with antigen stimulation, as was suggested before (Skurkovich et al. 1998). Possibly some groups of autoimmune diseases are connected with a special or defective type of IFN-γ or an altered IFN-γ pathway. There may also be cases in which cytokines are hypoproduced, which could also lead to disturbances in health. Every change in the production of IFN-γ and other cytokines can lead to pathology, but in particular to autoimmune conditions, including a temporary autoimmune condition. The main goal for rational treatment of Th-1-, Th-2-, and combined Th-1/Th-2-mediated diseases is to return the organism to homeostasis or a Th-1–Th-2 balance by suppressing or removing the hyperproduced cytokines with the help of antibodies, soluble receptors, vitamins, minerals and microelements, and different agents that suppress cytokine synthesis.

In conclusion, Th-1-mediated autoimmune diseases share certain characteristics. First, anti-IFN-γ or anti-TNF-α may generally be universal treatments for Th-1 autoimmune diseases, particularly skin diseases. Second, every Th-1 autoimmune disease develops through a pathological immune cascade. In general, removing one member of the cascade can have some therapeutic effect. Third, all autoimmune diseases develop with a reduced ability of the mononuclear cells to induce in vitro IFN-γ in response to stimulation. Fourth, deficiencies in natural killer cells are associated with autoimmune disease (Oikawa et al. 2003; Baxter and Smyth 2002). Fifth, in general autoimmune diseases are chronic, but in some cases can be temporary.

In autoimmune diseases, cytokines are often hyperproduced at the pathological site and then circulate in the blood, though finding cytokines in the blood is not always possible.

According to our investigations, anticytokine therapy holds great promise for treating many diseases, especially skin diseases. As described, we had good, sometimes striking, results with anti-IFN-γ and will continue testing this approach with other skin diseases. In our opinion, most skin diseases have a Th-1, Th-2, or Th-1/Th-2 genesis, and we feel our research, which introduced anticytokine therapy in 1974 (Skurkovich et al. 1974, 1989), will contribute to the understanding and treatment of these diseases. In this article, we have not discussed other cytokines besides IFN-γ, TNF-α, and IL-1 because of limited space and because we think these cytokines are most involved in the development of Th-1 autoimmune diseases.

References

Albers R, Bol M, Bleumink R, Willems AA, Pieters RH (2003) Effects of supplementation with vitamins A, C, and E, selenium, and zinc on immune function in a murine sensitization model. Nutrition 19:940–946

Arca E, Musabak U, Akar A, Erbil AH, Tastan HB (2004) Interferon-gamma in alopecia areata. Eur J Dermatol 14:33–36

Aurelian L, Ono F, Burnett J (2003) Herpes simplex virus (HSV)-associated erythema multiforme (HAEM): a viral disease with an autoimmune component. Dermatol Online J 9:1

Barker JN, Karabin GD, Stoof TJ et al (1991) Detection of interferon-gamma mRNA in psoriatic epidermis by polymerase chain reaction. J Dermatol Sci 2:106–111

Baxter AG and Smyth MJ (2002) The role of NK cells in autoimmune disease. Autoimmunity 35:1–14

Buntinx M, Ameloot M, Steels P (2002) Interferon-gamma-induced calcium influx in T lymphocytes of multiple sclerosis and rheumatoid arthritis patients: a complementary mechanism for T cell activation? J Neuroimmunol 124:70–82

Buntinx M, Gielen E, Van Hummelen P et al (2004) Cytokine-induced cell death in human oligodendroglial cell lines. II: alterations in gene expression induced by interferon-gamma and tumor necrosis factor-alpha. J Neurosci Res 76:846–861

Canete JD, Martinez SE, Farres J, Sanmarti R, Blay M, Gomez A, Salvador G, Munoz-Gomez J (2000) Differential Th-1/Th-2 cytokine patterns in chronic arthritis: interferon gamma is highly expressed in synovium of rheumatoid arthritis compared with seronegative spondyloarthropathies. Ann Rheum Dis 59:263–268

Cannon GW, Emkey RD, Denes A (1993) Prospective 5-year followup of recombinant interferon-gamma in rheumatoid arthritis. J Rheumatol 20:1867–1873

Centers for Disease Control and Prevention (CDC) (2004) Tuberculosis associated with blocking agents against tumor necrosis factor-alpha–California, 2002–2003. MMWR Morb Mortal Wkly Rep 53:683–686

Chinen J and Shearer WT (2003) Basic and clinical immunology. J Allergy Clin Immunol 111: S813–S818

Chopra V, Tyring SK, Johnson L, Fine JD (1992) Peripheral blood mononuclear cell subsets in patients with severe inherited forms of epidermolysis bullosa. Arch Dermatol 128:201–209

Christensen T, Sorensen PD, Hansen HJ, Moller-Larsen A (2001) [Endogenous retroviruses in multiple sclerosis] (in Danish) Ugeskr Laeger 163:297–301

Cippitelli M and Santoni A (1998) Vitamin D3: a transcriptional modulator of the interferon-gamma gene. Eur J Immunol 28:3017–3030

Cunnane G, Doran M, Bresnihan B (2003) Infections and biological therapy in rheumatoid arthritis. Best Pract Res Clin Rheumatol 17:345–363

Davis EG, Rush BR, Blecha F (2003) Increases in cytokine and antimicrobial peptide gene expression in horses by immunomodulation with Propionibacterium acnes. Vet Ther 4:5–11

Debray-Sachs M, Carnaud C, Boitard C, Cohen H, Gresser I, Bedossa P, Bach JF (1991) Prevention of diabetes in NOD mice treated with antibody to murine IFN gamma. J Autoimmun 4:237–248

Egwuagu CE, Mahdi RM, Chan CC, Sztein J, Li W, Smith JA, Chepelinsky AB (1999) Expression of interferon-gamma in the lens exacerbates anterior uveitis and induces retinal degenerative changes in transgenic Lewis rats. Clin Immunol 91:196–205

Ettefagh L, Nedorost S, Mirmirani P (2004) Alopecia areata in a patient using infliximab: new insights into the role of tumor necrosis factor on human hair follicles. Arch Dermatol 140:1012

Fabris P, Betterle C, Greggio NA, Zanchetta R, Bosi E, Biasin MR, de Lalla F (1998) Insulin-dependent diabetes mellitus during alpha-interferon therapy for chronic viral hepatitis. J Hepatol 28:514–517

Faergemann J, Jones JC, Hettler O, Loria Y (1996) Pityrosporum ovale (Malassezia furfur) as the causative agent of seborrhoeic dermatitis: new treatment options. Br J Dermatol 134 [Suppl 46]:12–5; discussion 38

Fierlbeck G, Rassner G, Muller C (1990) Psoriasis induced at the injection site of recombinant interferon gamma. Results of immunohistologic investigations. Arch Dermatol 126:351–355

Freyschmidt-Paul P, Zöller M, McElwee KJ, Sundberg J, Happle R, Hoffmann R (2004) The functional relevance of the type 1 cytokines IFN-γ and IL-2 in alopecia areata of C3H/HeJ mice. Plenary Workshop on Alopecia Areata, The 4th Intercontinental Meeting of Hair Research Societies, Berlin, Germany, June 17–19

Gardella R, Belletti L, Zoppi N, Marini D, Barlati S, Colombi M (1996) Identification of two splicing mutations in the collagen type VII gene (COL7A1)of a patient affected by the localisata variant of recessive dystrophic epidermolysis bullosa. Am J Hum Genet 59:292–300

Goodwin JL, Uemura E, Cunnick JE (1995) Microglial release of nitric oxide by the synergistic action of beta-amyloid and IFN-gamma. Brain Res 692:207–214

Grimes PE, Morris R, Avaniss-Aghajani E, Soriano T, Meraz M, Metzger A (2004) Topical tacrolimus therapy for vitiligo: therapeutic responses and skin messenger RNA expression of proinflammatory cytokines J Am Acad Dermatol 51:52–61

Hamilton CD (2004) Infectious complications of treatment with biologic agents. Curr Opin Rheumatol 16:393–398

Harrison LC and McColl GJ (1998) Infection and autoimmune disease In: Rose NR, Mackay IR (eds) The autoimmune diseases, 3rd edn. Academic Press, New York, 134

Hartung HP, Schafer B, van der Meide PH, Fierz W, Heininger K, Toyka KV (1990) The role of interferon-gamma in the pathogenesis of experimental autoimmune disease of the peripheral nervous system. Ann Neurol 27:247–257

Horwitz MS, Evans CF, McGavern DB, Rodriguez M, Oldstone MB (1997) Primary demyelination in transgenic mice expressing interferon-gamma. Nat Med 3:1037–1041

Jacob SE, Nassiri M, Kerdel FA, Vincek V (2003) Simultaneous measurement of multiple Th-1 and Th-2 serum cytokines in psoriasis and correlation with disease severity. Mediators Inflamm 12:309–313

Kageyama Y, Koide Y, Yoshida A et al (1998) Reduced susceptibility to collagen-induced arthritis in mice deficient in IFN-gamma receptor. J Immunol 161:1542–1548

Karlsson H, Bachmann S, Schroder J, McArthur J, Torrey EF, Yolken RH (2001) Retroviral RNA identified in the cerebrospinal fluids and brains of individuals with schizophrenia. Proc Natl Acad Sci U S A 98:4634–4639

Karpuzoglu-Sahin E, Hissong BD, Ansar Ahmed S (2001) Interferon-gamma levels are upregulated by 17-beta-estradiol and diethylstilbestrol. J Reprod Immunol 52:113–127

Kasparov AA, Narbut NP, Skurkovich, BS, Skurkovich SV (2004) Treatment of corneal transplant rejection using anti-interferon-γ and anti-tumor necrosis factor-alpha. In: Takhchidy KP (ed) New technology in the treatment of diseases of the cornea. Society of Ophthalmologists of the Russian Federation, Moscow, 200–202

Kawakami Y, Kuzuya N, Watanabe T, Uchiyama Y, Yamashita K (1990) Induction of experimental thyroiditis in mice by recombinant interferon gamma administration. Acta Endocrinol (Copenh)122:41–48

Khan A, Farah CS, Savage NW, Walsh LJ, Harbrow DJ, Sugerman PB (2003) Th-1 cytokines in oral lichen planus. J Oral Pathol Med 32:77–83

Koga T, Duan H, Urabe K, Furue M (2001) Immunohistochemical detection of interferon-gamma-producing cells in dermatophytosis. Eur J Dermatol 11:105–107

Koga T, Duan H, Urabe K, Furue M (2002) In situ localization of IFN-gamma-positive cells in psoriatic lesional epidermis. Eur J Dermatol 12:20–23

Kokuba H, Aurelian L, Burnett J (1999) Herpes simplex virus associated erythema multiforme (HAEM) is mechanistically distinct from drug-induced erythema multiforme: interferon-gamma is expressed in HAEM lesions and tumor necrosis factor-alpha in drug-induced erythema multiforme lesions. J Invest Dermatol 113:808–815

Korotky NG, Sharova NM, Skurkovich B, Skurkovich S (2004) Striking therapeutic effects in a patient with dysplastic epidermolysis bullosa using anti-IFN-γ (case report). IX International Congress of Dermatology, Beijing

Korotky NG, Sharova NM, Skurkovich B, Skurkovich S (2004) Successful anticytokine therapy in psoriasis. IX International Congress of Dermatology. Abstract Book, Beijing

Lenercept Multiple Sclerosis Study Group and The University of British Columbia MS/MRI Analysis Group (1999) TNF neutralization in MS: results of a randomized, placebo-controlled multicenter study. Neurology 53:457–465

Liu Q, Arseculeratne C, Liu Z, Whitmire J, Grusby MJ, Finkelman FD, Darling TN, Cheever AW, Swearengen J, Urban JF, Gause WC (2004) Simultaneous deficiency in CD28 and STAT6 results in chronic ectoparasite-induced inflammatory skin disease. Infect Immun 72:3706–715

Lukina GV (2004) Anticytokine therapy in rheumatoid arthritis. Summary of doctoral dissertation, Moscow. 23

Lukina G, Sigidin Y, Pushkova O, Mach E, Skurkovich B, Skurkovich S (2003) Treating rheumatoid arthritis with anti-interferon-γ and anti-tumor necrosis factor-α in a randomized, controlled trial. Eur Cytokine Netw 14 [September suppl]: 21

Mangge H, Gindl S, Kenzian H, Schauenstein K (2003) Atopic dermatitis as a side effect of anti-tumor necrosis factor-alpha therapy. J Rheumatol 30:2506–2507

Michel M, Duvoux C, Hezode C, Cherqui D (2003) Fulminant hepatitis after infliximab in a patient with hepatitis B virus treated for an adult onset Still's disease. J Rheumatol 30:1624–1625

Miossec P (1997) Cytokine-induced autoimmune disorders. Drug Saf 17:93–104

Molinero LL, Gruber M, Leoni J, Woscoff A, Zwirner NW (2003) Up-regulated expression of MICA and proinflammatory cytokines in skin biopsies from patients with seborrhoeic dermatitis. Clin Immunol 106:50–54

Moncada S, Palmer RM, Higgs EA (1991) Nitric oxide: physiology, pathophysiology, and pharmacology. Pharmacol Rev 43:109–142

Mosmann TR, Sad S (1996) The expanding universe of T-cell subsets: Th-1, Th-2 and more. Immunol Today 17:138–146

Mouser PE, Baker BS, Seaton ED et al (2003) Propionibacterium acnes-reactive T helper-1 cells in the skin of patients with acne vulgaris. J Invest Dermatol 121:1226–1228

Muller K, Bendtzen K (1996) 1,25-Dihydroxyvitamin D3 as a natural regulator of human immune functions. J Investig Dermatol Symp Proc 1:68–71

Murakami M, Kakizaki S, Takayama H, Takagi H, Mori M (1999) [Autoimmune thyroid disease induced by interferon therapy] (in Japanese) Nippon Rinsho 57:1779–1783

Nicoletti F, Zaccone P, Di Marco R, Lunetta M, Magro G, Grasso S, Meroni P, Garotta G (1997) Prevention of spontaneous autoimmune diabetes in diabetes-prone BB rats by prophylactic treatment with antirat interferon-gamma antibody. Endocrinology 138:281–288

Oikawa Y, Shimada A, Yamada S, Motohashi Y, Nakagawa Y, Irie J, Maruyama T, Saruta T (2003) NKT cell frequency in Japanese type 1 diabetes. Ann N Y Acad Sci 1005:230–232

Ozawa M and Aiba S (2004) Immunopathogenesis of psoriasis. Curr Drug Targets Inflamm Allergy 3:137–144

Panitch HS, Hirsch RL, Schindler J, Johnson KP (1987) Treatment of multiple sclerosis with gamma interferon: exacerbations associated with activation of the immune system. Neurology 37:1097–1102

Pittau E, Bogliolo A, Tinti A, Mela Q, Ibba G, Salis G, Perpignano G (1997) Development of arthritis and hypothyroidism during alpha-interferon therapy for chronic hepatitis C. Clin Exp Rheumatol 15:415–419

Ponsonby AL, McMichael A, van der Mei I (2002) Ultraviolet radiation and autoimmune disease: insights from epidemiological research. Toxicology 181/182:71–78

Prinz JC (1997) Psoriasis vulgaris, streptococci and the immune system: a riddle to be solved soon? Scand J Immunol 45:583–586

Reichrath J (2004) Antimycotics: why are they effective in the treatment of seborrheic dermatitis? Dermatology 208:174–175

Rocha IM and Guillo LA (2001) Lipopolysaccharide and cytokines induce nitric oxide synthase and produce nitric oxide in cultured normal human melanocytes. Arch Dermatol Res 293:245–48

Rufli T and Buchner SA (1984) T-cell subsets in acne rosacea lesions and the possible role of Demodex folliculorum. Dermatologica 169:1–5

Seitz M, Franke M, Kirchner H (1988) Induction of antinuclear antibodies in patients with rheumatoid arthritis receiving treatment with human recombinant interferon gamma. Ann Rheum Dis 47:642–644

Sigidin AY, Loukina GV, Skurkovich B, Skurkovich SV (2001) Randomized double-blind trial of anti-interferon-g antibodies in rheumatoid arthritis. Scan J Rheumatol 30:203–207

Sharova N, Korotky N, Skurkovich B, Skurkovich S (2003) Successful treatment of alopecia areata with anti-interferon (IFN)-γ. Proceedings of the 10th Meeting of the European Hair Research Society. 40

Singh R, Kumar A, Creery WD, Ruben M, Giulivi A, Diaz-Mitoma F (2003) Dysregulated expression of IFN-gamma and IL-10 and impaired IFN-gamma-mediated responses at different disease stages in patients with genital herpes simplex virus-2 infection. Clin Exp Immunol 133:97–107

Skurkovich SV, Klinova EG, Aleksandrovskaya IM, Levina NV, Arkhipova NA, Bulicheva TI (1973) Stimulation of transplantation immunity and plasma cell reaction by interferon in mice. Immunology 25:317–322

Skurkovich SV, Klinova EG, Eremkina EI, Levina NV (1974) Immunosuppressive effect of an anti-interferon serum. Nature 247:551–552

Skurkovich SV and Eremkina EI (1975) The probable role of interferon in allergy. Ann Allergy 35:356–360

Skurkovich SV, Skorikova AS, Dubrovina NA, Riabova TV, Eremkina EI, Golodnenkova VN et al (1977) Lymphocytes' cytotoxicity towards cells of human lymphoblastoid lines in patients with rheumatoid arthritis and systemic lupus erythematosus. Ann Allergy 39:344–350

Skurkovich S, Skurkovich B, Bellanti JA (1987) A unifying model of the immunoregulatory role of the interferon system: can interferon produce disease in humans? Clin Immunol Immunopathol 43:362–373

Skurkovich SV and Skurkovich B (1989) Development of autoimmune disease is connected with the initial disturbance of IFN synthesis in the cells. J IFN Res 9[suppl 2]:S305

Skurkovich S and Skurkovich B (1991) A new approach to the treatment of autoimmune diseases and AIDS. In: New American' Collected Scientific Reports. Buai Zion Soviet American Scientists Division, New York, 1:163–167

Skurkovich S, Skurkovich B, Bellanti JA (1993) A disturbance of interferon synthesis with the hyperproduction of unusual kinds of interferon can trigger autoimmune disease and play a pathogenetic role in AIDS: the removal of these interferons can be therapeutic. Med Hypoth 41:177–185

Skurkovich S, Skurkovich B, Bellanti JA (1994) A disturbance of interferon synthesis with the hyperproduction of unusual kinds of interferon can trigger autoimmune disease and play a pathogenetic role in AIDS: the removal of these interferons can be therapeutic. Med Hypoth 42:27–35

Skurkovich SV, Loukina GV, Sigidin YA, Skurkovich BS (1998) Successful first-time use of antibodies to interferon-gamma alone and combined with antibodies to tumor necrosis factor-alfa to treat rheumatic diseases (rheumatoid arthritis, systemic lupus erythematosus, psoriatic arthritis, Behcet's syndrome). Int J Immunother 14:23–32

Skurkovich SV, Boiko A, Beliaeva I et al (2001) Randomized study of antibodies to IFN-γ and TNF-α in secondary progressive multiple sclerosis. Multiple Sclerosis 7:277–284

Skurkovich S, Kasparov A, Narbut N, Skurkovich B (2002a) Treatment of corneal transplant rejection in humans with anti-interferon-gamma antibodies. Am J Ophthalmol 133:829–830

Skurkovich S, Korotky N, Sharova N, Skurkovich B (2002b) Anti-interferon-γ to treat alopecia areata, psoriasis vulgaris, vitiligo, pemphigus vulgaris, and epidermolysis bullosa. In You-Young Kim et al (eds) Proceedings of the 5th Asia Pacific Congress of Allergology and Clinical Immunology: 12–15 October 2002. Monduzzi Editore, 121–125

Skurkovich S, Korotky NG, Sharova NM, Skurkovich B. (2002c) Successful anti-IFN-γ therapy of alopecia areata, vitiligo, and psoriasis. Clin Immunol (Abstract supplement) 103:S103

Skurkovich SV, Skurkovich B, Kelly JA (2002d) Anticytokine therapy–new approach to the treatment of autoimmune and cytokine-disturbance diseases. Med Hypotheses 59:770–780

Skurkovich SV, Volkov IE, Delyagin VM, Rumyantzev AG, Skurkovich B (2003a) First use of anti-interferon (IFN)-γ in the treatment of type I diabetes in children. Berlin 8th International Workshop on Autoantibodies and Autoimmunity, Abstract book, poster 1

2 Cytokine Targeting in Psoriasis and Psoriatic Arthritis: Beyond TNFα

I.B. McInnes

2.1   Introduction   30
2.2   Interleukin-15   31
2.2.1   Basic Biology   31
2.2.2   Immunologic Activities of IL-15   33
2.3   IL-15 in Chronic Inflammation   35
2.4   IL-15 in Psoriasis and Psoriatic Arthritis   37
2.5   Conclusion   39
References   40

Abstract. Targeting TNFα provided proof of concept for the role of pro-inflammatory cytokines in promoting cutaneous inflammation, particularly psoriasis. Recent studies have elucidated the presence of numerous cytokine and chemokine activities in psoriatic skin and synovium. There is considerable interest in the potential of such activities as novel therapeutic targets. IL-15 is an innate response cytokine that activates leukocyte subsets via binding to its unique IL-15Rα and shared β and γ chain receptors. IL-15 promotes T cell memory and sustains local T cell activation, in part via prevention of apoptosis and mediates activation of monocytes, neutrophils and NK cells. IL-15 is up-regulated in psoriatic skin and psoriatic arthritis synovium. IL-15 blockade in a murine model of psoriasis led to marked suppression of typical psoriatic skin features. Clinical intervention in other chronic inflammatory disease states is now ongoing with encouraging early efficacy, raising the possibility for the first time of targeting this novel inflammatory moiety in psoriasis.

2.1 Introduction

Psoriasis is characterised by chronic cutaneous inflammation in up to 3% of the population. A variable proportion (8%–35%) of such patients develop articular involvement that in turn exhibits a clinical spectrum including dactyilits, synovitis, enthesitis, spondyloarthritis and arthritis mutilans. Topical and systemic immune modulatory agents used in various protocols form the current therapeutic state-of-the-art dependent upon disease severity and treatment resistance. The utility of therapeutic neutralisation of single cytokines in chronic inflammatory and autoimmune diseases has been unequivocally demonstrated in the success of biologic agents such as etanercept and infliximab. These agents target TNFα and have proven efficacy now across a range of diseases including rheumatoid arthritis, inflammatory bowel disease and uveitis (Taylor et al. 2001). Recent studies suggest that TNFα blockade is efficacious in psoriasis with remarkable responses detected via a number of clinical outcome measures in a substantial proportion of patients (Chaudhari et al. 2001; Leonardi et al. 2003). Similarly, improvements in articular disease in psoriatic arthritis is observed with etanercept and infliximab, and importantly the rate of articular destruction assessed by radiographic progression appears retarded, or arrested in some patients (Mease et al. 2000). Disease, however, recurs on discontinuation of therapy and partial or nonresponders remain problematic. In other inflammatory diseases, response rates may be improved with combination of methotrexate and/or treatment earlier in the disease course (Taylor et al. 2001; Klareskog et al. 2004). At present, it is uncertain whether combination therapies, revised dose schedules or indeed therapeutic strategies will be required to improve on existing response rates in psoriasis. It is clear, however, that there remains unmet clinical need. In particular, there is a critical requirement for agents that might not only suppress inflammation but also promote long-term recovery of peripheral tolerance and thereby disease remission, eventually achicvcd by 'drug-free' means.

Vital lessons have been learned in the therapeutic targeting of TNFα and of additional cytokines, particularly IL-1. Since single cytokine blockade can be successful despite the well-recognised functional overlap and pleiotropy of many described pro-inflammatory cytokines, this speaks to synergy whereby some cytokines may operate as 'keystones' or 'critical checkpoints' in the inflammatory cascade. Moreover, there appears to be an achievable therapeutic window in which critical immune suppression can be avoided whilst therapeutic response is achieved. A model has recently emerged that provides for rational choice of cytokines as targets. Cytokine expression should be examined in the tissue of interest in which functional bioactivities may be defined. Thereafter cytokines may be targeted in rodent models believed to reflect some component of human disease activity. Finally proof of concept trials may provide necessary data prior to continuing with appropriately powered placebo-controlled clinical trials. Pitfalls are evident in this approach, e.g. the relative disappointment of IL-1 blockade in humans relative to animal model studies. Nevertheless, this approach provides a useful basis upon which to discuss cytokine activities proposed as potential targets in chronic inflammatory diseases. This chapter will describe the basic biology of interleukin-15, an innate response cytokine that exhibits broad functional pleiotropy and that has been implicated in inflammatory dermatoses and malignancies. Thereafter it will describe current understanding of the role played by IL-15 in autoimmunity. Finally, it will describe our current understanding of the biology of IL-15 in the context of psoriatic inflammation in skin and joints.

2.2 Interleukin-15

2.2.1 Basic Biology

IL-15 was simultaneously described by Grabstein et al. and by Waldmann et al. in 1994. IL-15 (14–15 kDa) is a 4-α helix cytokine with structural similarities to IL-2 (Grabstein et al. 1994; Bamford et al. 1994). IL-15 mRNA is found in numerous normal human tissues and cell types, including activated monocytes, dendritic cells and fibroblasts (reviewed in Waldmann et al. 1999). Protein expression is more restricted, reflecting tight regulatory control of translation and secretion. IL-15 is subject to significant post-transcriptional regulation via 5′UTR AUG triplets, 3′ regulatory elements and a further putative C-terminus region regulatory site. Two isoforms of IL-15 with altered glycosylation are described – a long signalling peptide form (48-αα), which is secreted from the cell, and a short signalling peptide (21–αα) form, which remains intracellular, localised to nonendoplasmic regions in both cytoplasmic and nuclear compartments (Waldmann et al. 1999). Cell membrane expression may be crucial in mediating extracellular function rather than secretion and in part explains the difficulty in detecting soluble IL-15 in biologic systems.

IL-15 functions via a widely distributed heterotrimeric receptor (IL-15R) that consists of a β-chain (shared with IL-2) and common γ-chain, together with a unique α-chain (IL-15α) that in turn exists in eight isoforms (Waldmann 1999). This structure-function relationship has now been defined – IL-15 binds IL-15Rα via regions in the B helix and the C helix. A further region in the D-helix is involved in function but not α chain binding and presumably accounts for βγ chain interactions (Bernard et al. 2004). Of interest peptides generated in these studies offer potent inhibitory and agonistic IL-15 modulatory potential. The IL-15 receptor complex may form in *cis* or *trans* orientation allowing for receptor component sharing between adjacent cells (Dubois et al. 2002). Like IL-2, the IL-15Rαβγ complex signals through JAK1/3 and STAT3/5 (Giri et al. 1995; Waldmann et al. 1998). Additional signalling through *src*-related tyrosine kinases and Ras/Raf/MAPK to fos/jun activation is also proposed. High affinity ($10^{11}M^{-1}$) with slow off-rate makes IL-15Rα in soluble form a useful and specific inhibitor in biologic systems. Recently, a native soluble IL-15Rα has been identified that is present in plasma and is generated via proteolytic cleavage of membrane-expressed IL-15Rα by a matrix metalloproteinase (Mortier et al. 2004). Such cleavage occurs in part through activity of TACE (Budagian et al. 1994a). The natural sIL-15Rα has high affinity for native IL-15, and is capable of inhibiting IL-15 binding to membrane receptor, preventing effector cell function including proliferation. Whether such soluble IL-15Rα can contribute to receptor complex formation is unclear at this time.

Several intriguing recent observations suggest that IL-15 can function as a signalling molecule itself. Thus, membrane IL-15 exists as a membrane-bound moiety that is not dependent upon the presence of IL-15Rα. Such IL-15 was capable of promoting cytokine release in monocytes (Budagian et al. 2004b) and increased monocyte adhesion via the small molecular weight Rho-GTPase Rac3 (Neely et al. 2004). Several data suggest involvement also of MAPK pathways including erk1/2 and p38. Intriguing parallel analysis of the structure of IL-15LSP with TNFα reveals striking similarities in the presence of predicted transmembrane and cytoplasmic domains for both cytokines, suggesting that functional relationships for IL-15 may lie closer to TNFα than IL-2 (Budagian et al. 2004b).

2.2.2 Immunologic Activities of IL-15

Commensurate with the broad expression of IL-15R, diverse pro-inflammatory activities have been attributed to IL-15. Effects on natural killer (NK) and T cells are most prominent. IL-15 induces proliferation of mitogen-activated $CD4^+$ and $CD8^+$ T cells, T-cell clones and $\gamma\delta$ T cells, with release of soluble IL-2Rα, and enhances cytotoxicity both in $CD8^+$ T cells and lymphokine-activated killer cells (Grabstein et al. 1994; Bamford et al. 1994; Nishimura et al. 1996: Treiber-Held et al. 1996). CD69 expression is upregulated on $CD45RO^+$ but not $CD45RA^+$ T-cell subsets, consistent with the distribution of IL-2Rβ expression (Kanegane and Tosato 1996). IL-15 also induces NK cell activation, measured either by direct cytotoxicity, antibody-dependent cellular cytotoxicity or production of cytokines. It has been directly implicated in promoting NK cell mediated shock in mice. Moreover, IL-15 is implicated in thymic development of T cell and, particularly, NK cell lineages. IL-15 likely also functions as an NK cell survival factor in vivo by maintaining Bcl-2 expression (reviewed by Fehniger and Caligiuri 2001).

IL-15 exhibits T-cell chemokinetic activity and induces adhesion molecule (e.g. intercellular adhesion molecule [ICAM]-3) redistribution (Wilkinson and Liew 1995; Nieto et al. 1996). It further induces chemokine (CC-, CXC- and C-type) and chemokine receptor (CC but not CXC) expression on T cells (Perera and Waldmann 1999). Thus, IL-15 can recruit T cells and, thereafter, modify homo- or heterotypic cell–cell interactions within inflammatory sites. Whether IL-15 prejudices T helper 1 (Th1) or Th2 differentiation in addition to recruitment is controversial. IL-15 primes naive $CD4^+$ T cells from TCR-transgenic mice for subsequent IFNγ expression, but not IL-4 production (Seder 1996). Antigen-specific responses in T cells from human immunodeficiency virus-infected patients in the presence of high-dose IL-15 exhibit increased IFNγ production, particularly if IL-12 is relatively deficient (Seder et al. 1995). Similarly, IL-15 induces IFNγ/IL-4 ratios which favour Th1 dominance in mitogen-stimulated human T cells. However, IL-15 induces IL-5 production from allergen-specific human T-cell clones, implying a positive role in Th2-mediated allergic responses (Mori et al. 1996). Moreover, administration of soluble IL-15-IgG2b fusion protein in murine hypersensitivity models clearly implicates IL-15 in Th2 lesion development (Ruckert et al. 1998). Thus, through its function as a T-cell growth factor, IL-15 can probably sustain either Th1 or Th2 polarisation.

IL-15 mediates potent effects beyond T/NK cell biology. It promotes B-cell proliferation and immunoglobulin synthesis in vitro, in combination either with CD40 ligand (CD40L), or immobilised anti-IgM (Armitage et al. 1995). IL-15 has also been proposed as an autocrine regulator of macrophage activation, such that low levels of IL-15 suppress, whereas higher levels enhance pro-inflammatory monokine production (Alleva et al. 1997). Moreover, since human macrophages constitutively express bioactive membrane-bound IL-15, such autocrine effects are likely of early importance during macrophage activation (Musso et al. 1999). Similarly, neutrophils express IL-15Rα, and IL-15 can induce neutrophil activation, cytoskeletal rearrangement and protection from apoptosis (Girard et al. 1998). IL-15 has been proposed to promote angiogenesis and amplify fibroblast function via reversal of apoptosis. Finally, addition of IL-15 to rat bone marrow cultures induces osteoclast development and upregulates calcitonin receptor expression (Ogata et al. 1999).

IL-15 apparently represents a mechanism whereby host tissues can contribute to the early phase of immune responses, providing enhancement of polymorphonuclear and NK-cell responses, and subsequently T-cell responses, prior to optimal IL-2 production. The corollary to such pleiotropic activity may be a propensity to chronic, rather than self-limiting, inflammation should IL-15 synthesis be aberrantly regulated.

2.3 IL-15 in Chronic Inflammation

The foregoing description clearly renders IL-15 an intriguing candidate in the context of chronic inflammation. This has now been tested in a number of chronic inflammatory disease states (Table 1). Although expression data pertain to many of these diseases, most functional characterisation has been performed in inflammatory arthritis. We and other groups identified IL-15 and IL-15Rα expression at mRNA and protein levels in inflamed synovium and in peripheral blood from patients with RA (McInnes et al. 1997, 2003). Recently, IL-15 serum expression has been shown to rise progressively with RA disease duration (Gonzalez-Alvaro et al. 2003)]. IL-15 levels are also higher in juvenile idiopathic arthritis and correlate with CRP. IL-15 expression is also reported in RA nodule tissues together with a predominantly Th1 cytokine milieu (Hessian et al. 2003). It is clear therefore that IL-15 is expressed at the site of pathology. The functional importance of IL-15 in inflammatory synovitis is now established. IL-15 enhances synovial T cell proliferation, cytokine release, and optimises cognate interactions between T cells and macrophages that in turn lead to local monokine release including TNFα (McInnes et al. 2000). Additional effects on RA synovial fluid neutrophil activation, synovial NK cell granule release and fibroblast

Table 1. Targeting IL-15 in human disease

Effector function	Disease state
Enhance immune function	Human immunodeficiency virus
	Neoplastic disease
Suppress immune function	Transplantation
	Rheumatoid arthritis
	Psoriasis
	Psoriatic arthritis
	Inflammatory bowel disease
	Uveitis
	Sarcoidosis
Modify leukocyte growth	Lymphoma
	Haemopoietic disorders

apoptosis, proliferation and endothelial cell activation/migration have also been reported (reviewed in McInnes et al. 2003). High levels of serum cytokines including IL-15 correlate with T cell responses to type II collagen (CII) in vitro. In addition, CII-driven T cell/synovial fibroblast cross-talk amplified the release of IL-15, IL-17 and IFNγ via cell contact and CD40-dependent pathways (Cho et al. 2004). Moreover, IL-15 and TNF together promote expression of NKG2D on CD4$^+$ CD28$^-$ RA T cells that in turn were stimulated by synoviocyte expression of MIC ligands of NKG2D. IL-15 thereby opposes the effects of native regulators of this pathway and provides a potential route to enhanced autoreactivity (Groh et al. 2003).

Several approaches are in development to inhibit IL-15 in arthritis including neutralising monoclonal antibody, mutant IL-15:Fc fusion protein and soluble forms of the IL-15Rα (Table 2). A fully human antibody, HuMax-IL15 (Genmab A/C; antibody now termed AMG714) has been developed that is capable of binding not only soluble but also membrane-bound IL-15. AMG714 binds IL-15Rα-bound IL-15 and in vitro, effectively neutralises IL-15-mediated lymphocyte reversal of apoptosis and pro-inflammatory cytokine release. AMG714 is currently in phase I and II clinical trials in patients with active RA. Preliminary data indicate that this agent, administered by sc injection, is well tolerated in patients and when given weekly for 4 weeks produces encouraging ACR20 and ACR50 responses (Baslund et al., OP0008; EULAR 2003, *Ann Rheum Dis* 2003). However, the study design did not include a placebo agent throughout and so caution should be exercised in interpreting these data.

Table 2. Agents designed to modify IL-15 function

Enhance immune function	Recombinant IL-15
Suppress immune function	Anti-IL-15 (AMG714) antibody
	Soluble IL-15 receptor α
	IL-15:Fc mutant (CRB-15)
	Anti-IL-2/15 receptor β antibody
	Signalling molecular targets
	JAK3
	STAT3/5

In vivo targeting of IL-15 demonstrates that rodent CIA can be effectively ameliorated using soluble IL-15Rα. Treated DBA/1 mice exhibit reduced inflammation and suppression of histologic articular destruction (Ruchatz et al. 1998). We recently conducted a pilot study in collagen-induced arthritis in cynomolgus monkeys with sIL-15Rα and demonstrated reduction in CRP with suggestive effects on swollen joint count (unpublished data). This approach has not, however, been taken at this stage to human trials. A further intriguing possibility is the use of an IL-15:Fc fusion protein in which the IL-15 component is mutated to form a competitive antagonist. This moiety (designated CRB-15) suppresses delayed-type hypersensitivity in rodent models and delays cardiac transplant rejection (Kim et al. 1998; Ferrari-Lacraz et al. 2001). Thus, IL-15 clearly exhibits a suggestive bioactivity profile, is expressed in RA synovial tissues, is tractable in arthritis models and in early clinical trials shows promise as a target. It is clearly important to test the effect of IL-15 neutralisation in appropriately powered phase II clinical trials – such data will be available in the near future.

IL-15 function has been tested across a range of inflammatory conditions including pulmonary inflammation in which it is unlikely to represent a target in asthmatic conditions, but may offer promise in sarcoidosis or even COPD (reviewed in McInnes and Gracie 2004). Beneficial effects have been reported in models of diabetes and in transplantation across minor and major MHC incompatibility, in the latter utilised with either anti-CD4 or agonistic IL-2 and rapamycin (Zheng et al. 2003).

2.4 IL-15 in Psoriasis and Psoriatic Arthritis

IL-15 exhibits a plausible bioactivity profile in skin inflammation and has been detected in high levels in a range of inflammatory dermatoses. Thus, keratinocytes express both IL-15 and IL-15Rα, suggesting a role in autocrine regulation of keratinocytes in dermal inflammatory responses (Ruckert et al. 2000). Dermal IL-15 is upregulated by UVB irradiation and overexpression is associated with cutaneous malignancy (Mohamadzadeh et al. 1995). IL-15 has also been reported in bullous lesions, post-burn injury and in chronic cutaneous infection (Ameglio et al. 1999; Castagnoli et al. 1999). Whereas reduced levels are reported in atopic dermatitis that has been associated with propensity to enhanced Th2-mediated inflammation, increased IL-15 expression is consistently reported in psoriatic lesions (Ruckert et al. 2000; Ong et al. 2002). Psoriatic dermal pathology is characterised by local T cell, macrophage and neutrophil infiltration and activation, angiogenesis and epidermal hyperplasia, all features to which IL-15 may plausibly contribute. Prior studies associated dermal IL-15 with reduction in keratinocyte apoptosis together with promotion of T cell activation (Ruckert et al. 2000). Further insight as to the mechanisms whereby IL-15 mediates such effects emerged using the keratinocyte cell line HaCaT (Yano et al. 2003). IL-15-enhanced HaCaT proliferation was sensitive to addition of the MEK inhibitor U0126 or PI3-K inhibitor, LY294002. Erk1/2 and Akt phosphorylation, but not JNK, p38, STAT1 or STAT3, were observed in IL-15-treated epidermal cells and UVB-induced apoptosis was reversed. These data suggest that IL-15-mediated rescue from apoptosis previously observed in T cells could operate to promote epidermal proliferation in psoriasis. Complex interactions between T cells and keratinocytes mediated via cytokine networks including the Fas/FasL pathways have been proposed (Arnold et al. 1999). Similarly, direct IL-15-dependent interactions between skin fibroblasts and T cells have been proposed whereby TNF-exposed dermal fibroblasts upregulate membrane IL-15 to thereby activate adjacent T cells and suppress their apoptosis (Rappl et al. 2001). Thus TNFα and IL-15 cross-regulation of local inflammation is likely.

We extended these studies in an arthroscopic study of synovial membranes obtained prior to and after methotrexate therapy for 3 months. These studies showed that IL-15 mRNA and protein were modified but not abrogated by methotrexate treatment, raising the possibility that IL-15 targeting together with methotrexate might be reasonable (Kane et al., personal communication).

In vivo data suggesting a role for IL-15 first emerged in the psoriasis SCID model. A psoriatic patient derived CD94+/CD161+ T cell line injected into prepsoriatic skin engrafted on SCID mice induced development of psoriaform lesions characterised by local cytokine production. This cytokine network was of Th1 phenoytpe and predominated by IFNγ and IL-15 expression (Nickoloff et al. 2000). Definitive preclinical evidence for a pro-inflammatory role for IL-15 in psoriasis was subsequently derived using a human anti-IL-15 monoclonal antibody (146B7, Humax-IL15 above, now termed AMG714) raised in a human immunoglobulin transgenic in a xenograft murine model (Villadsen et al. 2003). AMG714 bound IL-15Rα that was in turn bound to IL-15 and in vitro, effectively neutralised IL-15-mediated lymphocyte reversal of apoptosis and pro-inflammatory cytokine release. Thereafter, human psoriatic skin (shown to express IL-15) was grafted onto SCID mice that in turn were treated with AMG714, cyclosporin or PBS. Significant reduction in histologic score, epidermal thickness, mononuclear infiltration, grade of parakeratosis and of keratinocyte cycling (via ki67 staining) were observed compared with either CyA or PBS-treated control grafts. These data are of importance for two reasons. They functionally implicate IL-15 in psoriatic pathogenesis. Moreover, they demonstrate for the first time the efficacy in vivo of antibody-mediated IL-15 neutralisation. Importantly, AMG714 apparently functions without interfering with IL-15:IL-15Rα interactions, operating through a nonclassical neutralisation pathway that allows blockade of the activities of prebound IL-15. This will theoretically facilitate interactions operating both on *cis*- and *trans*-configured IL-15R$\alpha\beta\gamma$ complexes (Dubois et al. 2002).

2.5 Conclusion

TNF blockade has provided substantial advances in therapeutic options and pathogenetic insight in psoriasis. Novel cytokine entities offer further clinical utility to address remaining unmet clinical need, either as single agents or in potential immune modulatory combinations. IL-15 is expressed in psoriatic skin and when targeted in vivo provides substantial improvement in skin inflammation. Antibodies that target IL-15 are in phase II clinical trials in RA and are under consideration in other inflammatory conditions. This therapy offers not only amelioration of inflammatory activity but also potential effects on T cell memory that may provide the potential for promoting immune tolerance in due course. This represents the ultimate goal of current immune modulatory strategies in the treatment of autoimmune diseases. Innovative trial designs will be required to test the latter possibility.

References

Alleva DG, Kaser SB, Monroy MA, Fenton MJ, Beller DI (1997) IL-15 functions as a potent autocrine regulator of macrophage proinflammatory cytokine production: evidence for differential receptor subunit utilization associated with stimulation or inhibition. J Immunol. 159:2941–2951

Ameglio F, D'Auria L, Cordiali-Fei P, Trento E, D'Agosto G, Mastroianni A, Giannetti A, Giacalone B (1999) Anti-intercellular substance antibody log titres are correlated with serum concentrations of interleukin-6, interleukin-15 and tumor necrosis factor-alpha in patients with Pemphigus vulgaris relationships with peripheral blood neutrophil counts, disease severity and duration and patients' age. J Biol Regul Homeost Agents 13:220–224

Armitage RJ, Macduff BM, Eisenman J, Paxton R, Grabstein KH (1995) IL-15 has stimulatory activity for the induction of B cell proliferation and differentiation. J Immunol 154:483–490

Arnold R, Seifert M, Asadullah K, Volk HD (1999) Crosstalk between keratinocytes and T lymphocytes via Fas/Fas ligand interaction: modulation by cytokines. J Immunol 162:7140–7147

Bamford RN, Grant AJ, Burton JD, Peters C, Kurys G, Goldman CK, Brennan J, Roessler E, Waldmann TA (1994) The interleukin (IL) 2 receptor beta chain is shared by IL-2 and a cytokine, provisionally designated IL-T, that stimulates T-cell proliferation and the induction of lymphokine-activated killer cells. Proc Natl Acad Sci U S A 91:4940–4944

Baslund B et al. (2004) OP0008; EULAR 2003, Ann Rheum Dis (in press)

Bernard J, Harb C, Mortier E, Quemener A, Meloen RH, Vermot-Desroches C, Wijdeness J, van Dijken P, Grotzinger J, Slootstra JW, Plet A, Jacques Y (2004) Identification of an interleukin-15alpha receptor-binding site on human interleukin-15. J Biol Chem 279:24313–24322

Budagian V, Bulanova E, Orinska Z, Ludwig A, Rose-John S, Saftig P, Borden EC, Bulfone-Paus S (2004a) Natural soluble interleukin-15Ralpha is generated by cleavage that involves the tumor necrosis factor-alpha-converting enzyme (TACE/ADAM17). J Biol Chem 279:40368–40375

Budagian V, Bulanova E, Orinska Z, Pohl T, Borden EC, Silverman R, Bulfone-Paus S (2004b) Reverse signaling through membrane-bound interleukin-15. J Biol Chem 279:42192–42201

Castagnoli C, Trombotto C, Ariotti S, Millesimo M, Ravarino D, Magliacani G, Ponzi AN, Stella M, Teich-Alasia S, Novelli F, Musso T (1999) Expression and role of IL-15 in post-burn hypertrophic scars. J Invest Dermatol 113:238–245

Chaudhari U, Romano P, Mulcahy LD, Dooley LT, Baker DG, Gottlieb AB (2001) Efficacy and safety of infliximab monotherapy for plaque-type psoriasis: a randomised trial. Lancet 357 (9271):1842–1847

Cho ML, Yoon CH, Hwang SY, Park MK, Min SY, Lee SH, Park SH, Kim HY (2004) Effector function of type II collagen-stimulated T cells from rheumatoid arthritis patients: cross-talk between T cells and synovial fibroblasts. Arthritis Rheum 50:776–784

Danning CL, Illei GG, Hitchon C, Greer MR, Boumpas DT, McInnes IB (2000) Macrophage-derived cytokine and nuclear factor kappaB p65 expression in synovial membrane and skin of patients with psoriatic arthritis. Arthritis Rheum 43:1244–1256

Dubois S, Mariner J, Waldmann TA, Tagaya Y (2002) IL-15Ralpha recycles and presents IL-15 In trans to neighboring cells. Immunity 17:537–547

Fehniger TA, Caligiuri MA (2001) Interleukin 15: biology and relevance to human disease. Blood 97:14–32

Ferrari-Lacraz S, Zheng XX, Kim YS, Li Y, Maslinski W, Li XC, Strom TB (2001) An antagonist IL-15/Fc protein prevents costimulation blockade-resistant rejection. J Immunol 167:3478–3485

Girard D, Boiani N, Beaulieu AD (1998) Human neutrophils express the interleukin-15 receptor alpha chain (IL-15Ralpha) but not the IL-9Ralpha component. Clin Immunol Immunopathol 88:232–240

Giri JG, Kumaki S, Ahdieh M, Friend DJ, Loomis A, Shanebeck K, DuBose R, Cosman D, Park LS, Anderson DM (1995) Identification and cloning of a novel IL-15 binding protein that is structurally related to the alpha chain of the IL-2 receptor. EMBO J 14:3654–3663

Gonzalez-Alvaro I, Ortiz AM, Garcia-Vicuna R, Balsa A, Pascual-Salcedo D, Laffon A (2003) Increased serum levels of interleukin-15 in rheumatoid arthritis with long- term disease. Clin Exp Rheumatol 21:639–642

Grabstein KH, Eisenman J, Shanebeck K, Rauch C, Srinivasan S, Fung V, Beers C, Richardson J, Schoenborn MA, Ahdieh M, Johnson L, Alderson MR, Watson JD, Anderson DM, Giri JG (1994) Cloning of a T cell growth factor that interacts with the beta chain of the interleukin-2 receptor. Science 264:965–968

Groh V, Bruhl A, El-Gabalawy H, Nelson JL, Spies T (2003) Stimulation of T cell autoreactivity by anomalous expression of NKG2D and its MIC ligands in rheumatoid arthritis. Proc Natl Acad Sci U S A 100:9452–9457

Hessian PA, Highton J, Kean A, Sun CK, Chin M (2003) Cytokine profile of the rheumatoid nodule suggests that it is a Th1 granuloma. Arthritis Rheum 48:334–338

Kanegane H, Tosato G (1996) Activation of naive and memory T cells by interleukin-15. Blood 88:230–235

Kim YS, Maslinski W, Zheng XX, Stevens AC, Li XC, Tesch GH, Kelley VR, Strom TB (1998) Targeting the IL-15 receptor with an antagonist IL-15 mutant/Fc gamma2a protein blocks delayed-type hypersensitivity. J Immunol 160:5742–5748

Klareskog L, van der Heijde D, de Jager JP, Gough A, Kalden J, Malaise M, Martin Mola E, Pavelka K, Sany J, Settas L, Wajdula J, Pedersen R, Fatenejad S, Sanda M (2004) Therapeutic effect of the combination of etanercept and methotrexate compared with each treatment alone in patients with rheumatoid arthritis: double-blind randomised controlled trial. Lancet 363:675–681

Leonardi CL, Powers JL, Matheson RT, Goffe BS, Zitnik R, Wang A, Gottlieb AB, Etanercept Psoriasis Study Group (2003) Etanercept as monotherapy in patients with psoriasis. N Engl J Med 349:2014–2022

McInnes IB, Gracie JA (2004) Interleukin-15: a new cytokine target for the treatment of inflammatory diseases. Curr Opin Pharmacol 392–397

McInnes IB, Leung BP, Sturrock RD, Field M, Liew FY (1997) Interleukin-15 mediates T cell-dependent regulation of tumor necrosis factor-alpha production in rheumatoid arthritis. Nat Med 3:189–195

McInnes IB, Leung BP, Liew FY (2000) Cell–cell interactions in synovitis. Interactions between T lymphocytes and synovial cells. Arthritis Res 2:374–378

McInnes IB, Gracie JA, Harnett M, Harnett W, Liew FY (2003) New strategies to control inflammatory synovitis: interleukin 15 and beyond. Ann Rheum Dis 62 [Suppl 2]:ii51–54

Mease PJ, Goffe BS, Metz J, VanderStoep A, Finck B, Burge DJ (2000) Etanercept in the treatment of psoriatic arthritis and psoriasis: a randomised trial. Lancet 356(9227):385–390

Mohamadzadeh M, Takashima A, Dougherty I, Knop J, Bergstresser PR, Cruz PD Jr (1995) Ultraviolet B radiation up-regulates the expression of IL-15 in human skin. J Immunol 155:4492–4496

Mori A, Suko M, Kaminuma O, Inoue S, Ohmura T, Nishizaki Y, Nagahori T, Asakura Y, Hoshino A, Okumura Y, Sato G, Ito K, Okudaira H (1996) IL-15 promotes cytokine production of human T helper cells. J Immunol 156:2400–2405

Mortier E, Bernard J, Plet A, Jacques Y (2004) Natural, proteolytic release of a soluble form of human IL-15 receptor alpha-chain that behaves as a specific, high affinity IL-15 antagonist. J Immunol 173:1681–1688

Musso T, Calosso L, Zucca M, Millesimo M, Ravarino D, Giovarelli M, Malavasi F, Ponzi AN, Paus R, Bulfone-Paus S (1999) Human monocytes constitutively express membrane-bound, biologically active, and interferon-gamma-upregulated interleukin-15. Blood 93:3531–3539

Neely GG, Epelman S, Ma LL, Colarusso P, Howlett CJ, Amankwah EK, McIntyre AC, Robbins SM, Mody CH (2004) Monocyte surface-bound IL-15 can function as an activating receptor and participate in reverse signaling. J Immunol 172:4225–4234

Nickoloff BJ, Bonish B, Huang BB, Porcelli SA (2000) Characterization of a T cell line bearing natural killer receptors and capable of creating psoriasis in a SCID mouse model system. J Dermatol Sci 24 :212–225

Nieto M, del Pozo MA, Sanchez-Madrid F (1996) Interleukin-15 induces adhesion receptor redistribution in T lymphocytes. Eur J Immunol 26:1302–1307

Nishimura H, Hiromatsu K, Kobayashi N, Grabstein KH, Paxton R, Sugamura K, Bluestone JA, Yoshikai Y (1996) IL-15 is a novel growth factor for murine gamma delta T cells induced by Salmonella infection. J Immunol 156:663–669

Ogata Y, Kukita A, Kukita T, Komine M, Miyahara A, Miyazaki S, Kohashi O(1999) A novel role of IL-15 in the development of osteoclasts inability to replace its activity with IL-2. J Immunol 162:2754–2760

Ong PY, Hamid QA, Travers JB, Strickland I, Al Kerithy M, Boguniewicz M, Leung DY (2002) Decreased IL-15 may contribute to elevated IgE and acute inflammation in atopic dermatitis. J Immunol 168):505–510

Perera L, Waldmann T (1999) IL-15 induces the expression of chemokines and their receptors in T lymphocytes. J Immunol 162:2606–2612

Rappl G, Kapsokefalou A, Heuser C, Rossler M, Ugurel S, Tilgen W, Reinhold U, Abken H (2001) Dermal fibroblasts sustain proliferation of activated T cells via membrane-bound interleukin-15 upon long-term stimulation with tumor necrosis factor-alpha. J Invest Dermatol 116):102–109

Ruchatz H, Leung BP, Wei XQ, McInnes IB, Liew FY (1998) Soluble IL-15 receptor alpha-chain administration prevents murine collagen-induced arthritis: a role for IL-15 in development of antigen-induced immunopathology. J Immunol 160:5654–5660

Ruckert R, Herz U, Paus R, Ungureanu D, Pohl T, Renz H, Bulfone-Paus S (1998) IL-15-IgG2b fusion protein accelerates and enhances a Th2 but not a Th1 immune response in vivo, while IL-2-IgG2b fusion protein inhibits both. Eur J Immunol 28:3312–3320

Ruckert R, Asadullah K, Seifert M, Budagian VM, Arnold R, Trombotto C, Paus R, Bulfone-Paus S (2000) Inhibition of keratinocyte apoptosis by IL-15: a new parameter in the pathogenesis of psoriasis? J Immunol 165:2240–2250

Seder RA, Grabstein KH, Berzofsky JA, McDyer JF (1995) Cytokine interactions in human immunodeficiency virus-infected individuals: roles of interleukin (IL)-2, IL-12, and IL-15. J Exp Med 182:1067–1077

Seder RA (1996) High-dose IL-2 and IL-15 enhance the in vitro priming of naive CD4[+] T cells for IFN-gamma but have differential effects on priming for IL-4. J Immunol 156:2413–2422

Taylor PC, Williams RO, Maini RN (2001) Immunotherapy for rheumatoid arthritis. Curr Opin Immunol 13:611–616

Treiber-Held S, Stewart DM, Kurman CC, Nelson DL (1996) IL-15 induces the release of soluble IL-2Ralpha from human peripheral blood mononuclear cells. Clin Immunol Immunopathol 79:71–78

Villadsen LS, Schuurman J, Beurskens F, Dam TN, Dagnaes-Hansen F, Skov L, Rygaard J, Voorhorst-Ogink MM, Gerritsen AF, van Dijk MA, Parren PW, Baadsgaard O, van de Winkel JG (2003) Resolution of psoriasis upon blockade of IL-15 biological activity in a xenograft mouse model. J Clin Invest 112:1571–1580

Waldmann TA, Tagaya Y (1999) The multifaceted regulation of interleukin-15 expression and the role of this cytokine in NK cell differentiation and host response to intracellular pathogens. Annu Rev Immunol 17:19–49

Waldmann T, Tagaya Y, Bamford R (1998) Interleukin-2, interleukin-15, and their receptors. Int Rev Immunol 16:205–226

Wilkinson PC, Liew FY (1995) Chemoattraction of human blood T lymphocytes by interleukin-15. J Exp Med 181:1255–1259

Yano S, Komine M, Fujimoto M, Okochi H, Tamaki K (2003) Interleukin 15 induces the signals of epidermal proliferation through ERK and PI 3-kinase in a human epidermal keratinocyte cell line, HaCaT. Biochem Biophys Res Commun 301:841–850

Zheng XX, Sanchez-Fueyo A, Sho M, Domenig C, Sayegh MH, Strom TB (2003) Favorably tipping the balance between cytopathic and regulatory T cells to create transplantation tolerance. Immunity 19:503–514

3 Inhibitors of Histone Deacetylases as Anti-inflammatory Drugs

C.A. Dinarello

3.1	Introduction	46
3.1.1	HDAC Inhibitors as Anti-tumor Agents	47
3.1.2	HDAC Inhibition and Expression of Latent Viral Genes	47
3.2	HDAC Inhibitors in Models of Inflammatory Diseases	48
3.2.1	Models of Lupus Erythematosus	48
3.2.2	Graft-Versus-Host Disease	50
3.2.3	Models of Arthritis	50
3.2.4	Hepatitis Induced by Intravenous Con A	51
3.3	Reducing Cytokines by HDAC Inhibitors	51
3.3.1	Low Concentrations of HDAC Inhibitors are Anti-inflammatory Whereas High Concentrations Are Needed for Anti-tumor Effects	51
3.3.2	Effect of SAHA on Secretion of IL-1β	52
3.3.3	Other Cytokines	53
3.3.4	LPS-Induced Cytokines In Vivo	54
3.4	Mechanism of Action of HDAC Inhibition in Reducing Inflammation	55
References		56

Abstract. This review addresses the issue of histone deacetylase (HDAC) inhibitors as developed for the treatment of cancer and for the investigation of the inhibition of inflammation. The review focuses on both in vitro and in vivo models of inflammation and autoimmunity. Of particular interest is the inhibition of pro-inflammatory cytokines. Although the reduction in cytokines appears paradoxical at first, upon examination, some genes that are anti-inflammatory are upregulated by inhibition of HDAC. Whether skin diseases will be affected by inhibitors of HDAC remains to be tested.

3.1 Introduction

In order to maintain the compact nature of DNA, chromatin is tightly wrapped around nuclear histones in distinct units called nucleosomes. The enzyme family called histone deacetylases (HDAC) maintains the histone proteins in a state of deacetylation so that DNA can bind tightly. A natural balance exists between histone acetylases (HAC) and HDAC. Synthetic inhibitors of histone deacetylases result in hyperacetylation of histones and the unraveling of the chromatin tightly wrapped in the nucleosome and allow transcription factors to bind and initiate gene expression. Developed for the treatment of cancer, inhibitors of HDAC increase the expression of a variety of genes, which are silenced in malignant cells. As such, the anti-tumor effects of HDAC inhibitors increase the expression of genes driving cell cycle, tumor suppression, differentiation and apoptosis (Marks et al. 2000, 2001; Richon et al., 1998, 2000, 2001). Suberoylanilide hydroxamic acid (SAHA) belongs to the class of hydroxamic acid-containing hybrid polar molecules that inhibit HDAC. SAHA suppresses the proliferation of cancer cells in vitro and reduces the growth of experimental tumors in vivo (Butler et al. 2000; Marks et al. 2001; Richon et al. 2001). SAHA, trichostatin A and butyrate are well-studied inhibitors of nuclear HDAC. However, SAHA also binds to S3 protein in the cytosol, a component of the ribosome (Webb et al. 1999). There are ongoing clinical trials of SAHA (Marks et al. 2001), and patients with cancer have been injected with increasing doses of SAHA (300–600 mg/m^2) intravenously (O'Connor et al. 2001). Although solid tumors are treated in clinical trails with HDAC inhibitors, leukemias and multiple myeloma are often cancers that are first studied for treatment with HDAC inhibitors.

3.1.1 HDAC Inhibitors as Anti-tumor Agents

A large number of studies have revealed that inhibitors of HDAC reduce the proliferation of transformed cells in vitro as well as the growth of experimental tumors in vivo. The HDAC inhibitor depsipeptide has been administered to three patients with cutaneous T-cell lymphoma associated with clinical responses (Pierart et al. 1988). SAHA has also advanced to clinical studies in patients with prostatic cancer and lymphoma. The physical property of SAHA to inhibit HDAC is its binding of the hydroxamic acid moiety to the zinc-containing pocket of HDAC (Finnin et al. 1999). This results in increased acetylated histones. SAHA inhibits HDAC 1 and 3 and there is hyperacetylation of histones 3 and 4 (Richon et al. 1998). As a consequence of hyperacetylation, SAHA and other inhibitors of HDAC increase the expression of approximately 1%–2% of genes (Marks et al. 2000). Increased gene expression for the cell cycle kinase inhibitor p21 (Richon et al. 2000), as well as other mechanisms of tumor cell apoptosis, account for the anti-tumor properties of SAHA (Said et al. 2001; Vrana et al. 1999). In vitro, micromolar concentrations of SAHA result in selective apoptotic cell death, terminal differentiation and growth arrest for tumor cells, without toxicity on normal cells. In mice, growth of transplanted human prostatic cancer cells was suppressed by 97% by daily administration of SAHA at 50 mg/kg per day (Butler et al. 2000).

3.1.2 HDAC Inhibition and Expression of Latent Viral Genes

In addition to suppressing tumor growth, inhibitors of HDAC may affect other regulatory pathways. Trichostatin, and other inhibitors of HDAC, for example, increases viral expression, particularly latent viral genes. However, the most commonly used HDAC inhibitor, butyrate, is not associated with increased expression of Herpes viruses. Some studies in vitro suggest that HDAC inhibitors can be used to express genes in vectors used for gene therapy. However, the greatest interest in the therapeutic use of HDAC inhibitors for increased expression of viral genes is that of HIV-1 (Demonte et al. 2004; Hsia and Shi 2002; Van Lint et al. 1996). Since the onset of highly active anti-retroviral therapy for HIV-1, suppression of HIV-1 infection has resulted in a remarkable prolongation of life and a return, in part, of $CD4^+$ T cells.

Despite the near absence of HIV-1 mRNA in the serum and even the inability to detect HIV-1 in peripheral blood mononuclear cells (PBMCs) and even tonsillar tissue, withdrawal of anti-retroviral therapy precipitates a near immediate return of viral mRNA into the circulation. It has become clear that latent virus exists in a cellular compartment not accessible to therapy. Forcing the expression of HIV-1 incorporated into genomic DNA in sequestered cells serves as a reservoir of the infection. In order to "force" expression of HIV-1, which results in the death of the infected cell, cytokines such as IL-2 have been given to patients while on anti-retroviral therapy. The use of cytokines to "force" expression and to increase HIV-1 expression depends upon signal transduction following cell surface cytokine receptor activation. This mechanism is thought to be useful in "purging" HIV-1-infected cells to express latent virus from the host. But this approach has failed. One explanation for the failure of cytokine therapy to purge HIV-1 infection is that the reservoir of latently infected cells may not express the receptor for the particular cytokine. For example, it is unlikely that epithelial cells express receptors for IL-2. The advantage of HDAC inhibitors for purging HIV-1 is twofold: these inhibitors are small molecules and may have access to cellular compartments not accessible to cytokines. Second, HDAC inhibitors enter cells via a cell surface receptor-independent mechanism. HDAC inhibition may be a useful in HIV-1 treatment since the agents can be administered in cycles with anti-retroviral agents given immediately after the purge to prevent infection of new cells.

3.2 HDAC Inhibitors in Models of Inflammatory Diseases

3.2.1 Models of Lupus Erythematosus

An unexpected finding of HDAC inhibitors was the reduction in disease severity of models of murine autoimmune disease. The mouse model for systemic lupus erythematosus is the lpr/lpr mouse that develops a spontaneous disease characterized by nephritis, proteinuria and early death. Trichostatin A was injected into these mice before the onset of significant disease for 5 weeks. Because trichostatin A in water is insoluble, the vehicle was also used for 5 weeks of treatment. Trichostatin A treatment resulted in significantly less proteinuria; in addition, there was histological evidence of decreased glomerulonephritis (Mishra et al. 2003), as well as a reduction in spleen weight. A concomitant decreased expression of steady state levels of mRNA for IL-12, IFNγ, IL-6, and IL-10 was observed as well as protein levels for these cytokines. Not unexpected, total cellular chromatin contained an accumulation of acetylated histones H3 and H4.

These mouse studies were consistent with a similar effect of trichostatin A in peripheral T cells from patients with systemic lupus erythematosus (Mishra et al. 2001). In T-helper cells from patients with the disease, there is overexpression of the T-helper 2 cytokine IL-10 and the pro-inflammatory cytokine CD40 ligand (also known as CD154). However, incubation of the cells from these patients with trichostatin A decreased spontaneous IL-10 and CD40 ligand. As a consequence of decreased IL-10, there was an increase in IFNγ production. Further studies in the lpr/lpr mouse model were carried out using the water-soluble SAHA compound. SAHA was injected into lpr/lpr mice from age 10–20 weeks. Although this treatment did not affect the level of auto-antibodies that develop in these mice and although the deposition of anti-glomerular immunoglobulin was unaffected by SAHA treatment, there was a marked reduction in the histological abnormalities of the kidneys and a decrease in the proteinura (Reilly et al. 2004). In addition, spleen weight was decreased as well as fewer CD4-CD8 cells, both pathological indicators for the disease. Mesangial cells from mice treated with SAHA, when stimulated with LPS plus IFNγ, produced less spontaneous nitric oxide (Reilly et al. 2004).

IFNγ is an important cytokine in several autoimmune diseases but is of particular importance in the pathogenesis of lupus erythematosus. The animal model for this disease is the lpr/lpr mouse, which develops a spontaneous proteinuria and lethal nephritis. Neutralization of IL-18 reduced the proteinuria and decreased the lethality (Bossu et al. 2003). In human PBMCs, SAHA reduces IFNγ induced by endotoxin or by the combination of IL-12 plus IL-18 (Leoni et al. 2002). Although there was an increase in IFNγ production in T cells from lupus patients exposed to trichostatin (Mishra et al. 2001), the effect was most likely due to a reduction in IL-10.

3.2.2   Graft-Versus-Host Disease

Relevant to the issue of HDAC inhibition of IFNγ production is the mouse model of graft-versus-host disease (GVHD), a model for GVHD in humans following bone marrow transplantation. GVHD and leukemic relapse are the two major obstacles to successful outcomes after allogeneic bone marrow transplantation. Human and mouse studies have demonstrated dysregulation of proinflammatory cytokines with the loss of gastrointestinal tract integrity contributing to GVHD. In mice with acute GVHD, the administration of SAHA at low doses reduced IFNγ, TNFα, IL-1β and IL-6 production (Reddy et al. 2004). In addition, intestinal histopathology, clinical severity, and mortality were reduced compared with vehicle-treated animals (Reddy et al. 2004). However, SAHA did not impair graft versus leukemia responses and significantly improved leukemia-free survival by using two different tumors (Reddy et al. 2004). These findings are consistent with the ability of SAHA to suppress LPS-induced TNFα and IFNγ as well as IL-12/IL-18-induced IFNγ and IL-6, but also with the failure of SAHA to reduce IFNγ and other cytokines stimulated by anti-CD3 agonistic antibodies.

3.2.3   Models of Arthritis

The therapeutic benefit of blocking pro-inflammatory cytokines in the progression of rheumatoid arthritis has been based on models of the disease in mice and rats. Two models have been used: collagen-induced arthritis and adjuvant arthritis. In the latter, the pathological effects are the destruction of bone and cartilage. The role of cytokines in this process is well established. For example, IL-1β and IL-18 are potent inducers of metalloproteinases and inhibitors of proteoglycan synthesis (Bakker et al. 1997; Joosten et al. 2003; van den Berg et al. 1994). Using the model of rat adjuvant arthritis, treatment with phenylbutyrate or trichostatin A resulted in decreased expression of TNFα (Chung et al. 2003). This was associated with a decrease in joint swelling, synovial mononuclear cell infiltration, synovial hyperplasia, and pannus formation. There was also a remarkable absence of cartilage or bone destruction, commonly observed in adjuvant arthritis (Chung et al. 2003).

The HDAC inhibitor depsipeptide (FK228) was used to treat mice using a model of rheumatoid arthritis induced by autoantibody. Following the establishment of the arthritis, mice were injected intravenously with 2.5 mg/kg and clinical scores were obtained. A single dose injection of FK228 reduced joint swelling, synovial cell infiltration and also decreased bone and cartilage destruction (Nishida et al. 2004). There was a reduction in TNFα and IL-1β production associated with histone hyperacetylation in the synovial cells. However, at the doses used, FK228 inhibited the in vitro proliferation of synovial fibroblasts, by a mechanism that may result in arrest of cell cycle. Indeed, FK228 increased acetylation of p21 in synovial cells (Nishida et al. 2004). An important concept is that inhibitors of HDAC at high doses act as anti-proliferative agents and must be distinguished from a purely anti-cytokine effect.

3.2.4 Hepatitis Induced by Intravenous Con A

The in vivo model of ConA-induced hepatic injury, a model that is TNFα- and IL-18-dependent, is a model of activated CD4$^+$ T cells that may represent autoimmune hepatitis and possibly most cytokine-mediated hepatic injury (Faggioni et al. 2000; Fantuzzi et al. 2003). In this model, a single dose of 50 mg/kg SAHA given orally reduced the injury by more than 70% as determined by a reduction in circulating levels of ALT by (Leoni et al. 2002).

3.3 Reducing Cytokines by HDAC Inhibitors

3.3.1 Low Concentrations of HDAC Inhibitors are Anti-inflammatory Whereas High Concentrations are Needed for Anti-tumor Effects

It is interesting to note that the inhibitory effect of SAHA appears to be particularly effective for reducing proinflammatory cytokines at relatively lower doses compared to the doses used for inhibition of tumors. In fact, the in vitro concentrations of SAHA that inhibited proliferation of tumor cell lines by 50% were 1–5 μM (Leoni et al. 2002), similar to those described by others (Butler et al. 2000). However, at nanomolar concentrations (50–200 nM), a 50%–85% reductions in LPS-induced secretion of TNFα, IL-1β, IFNγ, and IL-12 in freshly isolated human PBMCs was reported (Leoni et al. 2002). Similar to the reduction in protein levels of TNFα and IFNγ, the steady state mRNA levels for these two cytokines were reduced, particularly those of IFNγ. In addition, the in vitro production of nitric oxide in primary mouse peritoneal macrophages stimulated with the combination of TNFα and IFNγ was suppressed by SAHA at 200–400 nM (Leoni et al. 2002). Therefore, the anti-inflammatory effect of SAHA is evident at concentrations lower than those needed to suppress tumor cell growth in vitro and in vivo.

IFNγ production triggered by the T-cell receptor using anti-CD3 was unaffected by SAHA (Leoni et al. 2002). In contrast, when stimulated by either LPS or the combination of IL-18 plus IL-12, IFNγ production was markedly reduced by SAHA. The intracellular and extracellular levels of the anti-inflammatory cytokine IL-1 receptor antagonist were unaffected by SAHA in LPS-stimulated human PBMCs and there was no reduction in circulating IL-10 levels in mice injected with LPS. SAHA also did not reduce the production of the proinflammatory chemokine IL-8. Consistent with this finding, steady-state mRNA for IL-8 was unaffected by SAHA (Leoni et al. 2002).

3.3.2   Effect of SAHA on Secretion of IL-1β

Pretreatment with SAHA resulted in reduced secretion of mature IL-1β from PBMCs and the level of circulating IL-1β in LPS-injected mice (Leoni et al. 2002). However, SAHA did not affect steady state levels of IL-1β mRNA or the intracellular levels of precursor IL-1β. In the same PBMC cultures, SAHA reduced both the secretion of TNFα and IFNγ as well as their mRNA levels. Therefore, the reduction in IL-1β by SAHA appears to be primarily at the level of secretion of mature IL-1β. Although the pathway(s) for secretion of IL-1β remains unclear, there is no dearth of evidence that inhibition of caspase-1 results in decreased secretion of mature IL-1β, particularly in LPS-stimulated human monocytes (reviewed in Dinarello 1996). When tested, SAHA at concentrations as high as 10 μM did not inhibit caspase-1 enzymatic activity. However, SAHA may increase gene expression and synthesis of the intracellular inhibitor of ICE, serpin proteinase inhibitor 9 (Young et al. 2000).

3.3.3   Other Cytokines

Butyrate reduces IL-12 production but upregulates IL-10 in human blood monocytes (Saemann et al. 2000), although millimolar concentrations are needed. Butyrate also inhibits the proliferation of the Caco-2 colon cancer epithelial cells and reduces spontaneous gene expression of IL-8 in these cells (Huang et al. 1997). Using the same cell line, others have reported that butyrate or trichostatin increase IL-8 production (Fusunyan et al. 1999). It is not uncommon for trichostatin and butyrate to have paradoxical effects on IL-8 production by Caco-2 cells; this is likely due to the state of cell activation. For example, IL-8 secretion was increased by 145% by trichostatin and butyrate alone, whereas in TNFα-stimulated cells IL-8 was inhibited to levels below unstimulated levels (Gibson et al. 1999).

In PBMCs from healthy subjects stimulated with LPS or the combination of IL-12/IL-18 in vitro, there is a consistently observed marked reduction in gene expression and synthesis of IFNγ by SAHA. Other investigators have reported that at apoptosis-inducing concentrations of trichostatin and butyrate, IFNγ production in T lymphocytes is reduced (Dangond and Gullans 1998). In addition, butyrate or trichostatin inhibited IL-1β-dependent induction of the acute phase protein gene haptoglobin as well as the binding of transcription factors to the haptoglobin promoter (Desilets et al. 2000). Butyrate and trichostatin also suppress IL-2 induction of c-myc, bag-1, and LC-PTP gene expression (Koyama et al. 2000). The IL-2 enhancer and promoter were suppressed by trichostatin by 50% at 73 nM, whereas at the same concentration there was increased expression directed by the c-fos enhancer and promoter (Takahashi et al. 1996).

Most in vitro studies on inhibitors of HDAC have been carried out using cell lines, particularly cell lines of malignant origin. In contrast, the effects of HDAC inhibitors on primary cells or in vivo studies are more likely to be applied to clinical uses of the agents to treat inflammatory and autoimmune diseases. Diseases of auto-immune/inflammatory nature have altered gene expression profiles as compared to health, and thus one must look at the effect of HDAC inhibitors in stimulated cells rather than resting cells. In the absence of exogenous stimuli, SAHA has no effect on the production of cytokines and steady-state mRNA levels of cytokines are unaffected (Leoni et al. 2002). In contrast, LPS-induced TNFα and IFNγ mRNA were decreased by treatment with SAHA. Therefore, the anti-inflammatory effects of SAHA and possibly other inhibitors of HDAC is selective and likely dependent on the level of activation, particularly in primary cells. Alternatively, SAHA may have anti-inflammatory properties independent of its ability to inhibit nuclear HDAC, for example, hyperacetylation of nonhistone proteins such as ribosomal S3 or the Rel A subunit of NFκB (Takahashi et al. 1996).

3.3.4 LPS-Induced Cytokines In Vivo

LPS-induced circulating TNFα and IL-1β, as well as IFNγ, were suppressed by SAHA (Leoni et al. 2002). The use of LPS as an inducer of cytokines is widely accepted as a model of disease. However, with the exception of inflammatory bowel disease, most cytokine-mediated autoimmune diseases are triggered by nonmicrobial products such as autoantibodies and several endogenous cytokines themselves, particularly CD4$^+$ T cell products. IL-12 and IL-18 are primarily macrophage products, which in turn stimulate T lymphocytes to produce IFNγ and IL-6. SAHA reduces both LPS- as well as IL-12/IL-18-induced IFNγ and IL-6. IL-6 is an important mediator of inflammation, primarily as a B-cell growth factor and an inducer of hepatic acute-phase protein synthesis in diseases such a multiple myeloma (Lust and Donovan 1999), and antibodies to the IL-6 receptor are used to treat multiple myeloma, rheumatoid arthritis, and other autoimmune diseases (Iwamoto et al. 2002). Gene expression for IFNγ in resting PBMCs stimulated with LPS was nearly completely absent 24 h after treatment with SAHA. However, either in vitro or in vivo, direct stimulation of T-cell receptor using agonistic anti-CD3-induced IFNγ was unaffected by SAHA (Leoni et al. 2002). It appears that IL-12/IL-18 induces IFNγ via a pathway, which is sensitive to inhibition by SAHA, whereas the pathway for IFNγ via the T-cell receptor is not.

3.4 Mechanism of Action of HDAC Inhibition in Reducing Inflammation

It is important to distinguish between the effects of HDAC inhibition on cytokine production in transformed cell lines with those in primary cells or effects in vivo. Of the several genes coding for the different isoforms of HDAC, HDAC3, a Class I deacetylase, has been studied for its effects on the transcription of TNFα. Overexpression of HDAC3 reduced the ability of the p65 component of NFκB to transcribe the TNFα gene (Miao et al. 2004), whereas inhibition of HDAC, in this particular model, stimulated TNFα expression. Others have reported that HDAC3 associates with a nonhistone substrate, acetylated RelA (p65) and deacetylates the molecule (Chen et al. 2001). In doing so, acetylation/deacetylation of p65 controls the transcriptional responses by affecting the binding to inhibitory κB (Chen et al. 2001). Nonhistone interaction of HDAC3 has been described with mitogen activated protein kinase (MAPK) 11 (Mahlknecht et al. 2004), in which the deacetylation of this kinase prevents activation of MAPK11 (also known as MAPK p38) participation in LPS-induced gene expression. Inhibition of phosphorylation of MAPK11 prevents transcription of TNFα and other proinflammatory cytokines by blocking the binding of activating transcription factor-2 (ATF-2). It was demonstrated that overexpression of HDAC3 suppresses LPS-induction of TNF gene expression (Mahlknecht et al. 2004). Not unexpectedly, it was shown that inhibition of trichostatin A overcomes the deacetylation by HDAC3 and increases the production of TNFα in LPS-stimulated macrophage cell lines (Mahlknecht et al. 2004).

Also in macrophage cell lines, TNFα production is increased upon inhibition of HDAC (Lee et al. 2003) due to acetylation of H3 and H4. In primary monocytes, it is only upon maturation during 7 days of culture that a similar acetylation takes place and increased expression of TNFα is observed (Lee et al. 2003). The inhibition of TNFα production by p50 homodimers binding to the TNFα promoters is well established. However, this inhibition is reversed by HDAC inhibition and increased TNFα is observed (Wessells et al. 2004). In order to resolve the discrepancies between increased or decreased TNFα production by inhibition of HDAC, one may conclude that in cell lines increases in TNFα production are consistently observed and may be due to acetylation of cellular proteins or nuclear histone. On the other hand, inhibition of TNFα gene expression or synthesis by SAHA is consistently observed in primary cells in vitro and in vivo models of disease.

Clearly, from the present studies and also from previous reports (Mishra et al. 2003), inhibition of HDAC3 is not a likely target of SAHA at concentrations that suppress cytokines. The treatment of lupus-prone mice with trichostatin A reduced proteinuria, the infiltration of destructive inflammatory cells into the glomerulus, and reduced spleen weight (Mishra et al. 2003). The clinical benefit of trichostatin A in these mice was associated with decreases in IL-12, IFNγ, and IL-6. The evidence that SAHA primarily suppresses inflammation in animal models of disease (Leoni et al. 2002; Mishra et al. 2003; Reddy et al. 2001; Reilly et al. 2004) via suppression of TNFα, IL-1β, IFNγ, and IL-6 transcription is consistent with a histone rather than a nonhistone target for hyperacetylation in primary cells.

Acknowledgements. These studies are supported by NIH Grants AI-15614 and HL-68743 and the Colorado Cancer Center

References

Bakker AC, Joosten LA, Arntz OJ, Helsen MM, Bendele AM, van de Loo FA, van den Berg WB (1997) Prevention of murine collagen-induced arthritis in the knee and ipsilateral paw by local expression of human interleukin-1 receptor antagonist protein in the knee. Arthritis Rheum 40:893–900

Bossu P, Neumann D, Del Giudice E, Ciaramella A, Gloaguen I, Fantuzzi G, Dinarello CA, Di Carlo E, Musiani P, Meroni PL, Caselli G, Ruggiero P, Boraschi D (2003) IL-18 cDNA vaccination protects mice from spontaneous lupus-like autoimmune disease. Proc Natl Acad Sci U S A 100:14181–14186

Butler LM, Agus DB, Scher HI, Higgins B, Rose A, Cordon-Cardo C, Thaler HT, Rifkind RA, Marks PA, Richon VM (2000) Suberoylanilide hydroxamic acid, an inhibitor of histone deacetylase, suppresses the growth of prostate cancer cells in vitro and in vivo. Cancer Res 60:5165–1570

Chen L, Fischle W, Verdin E, Greene WC (2001) Duration of nuclear NF-kappaB action regulated by reversible acetylation. Science 293:1653–1657

Chung YL, Lee MY, Wang AJ, Yao LF (2003) A therapeutic strategy uses histone deacetylase inhibitors to modulate the expression of genes involved in the pathogenesis of rheumatoid arthritis. Mol Ther 8:707–717

Dangond F, Gullans SR (1998) Differential expression of human histone deacetylase mRNAs in response to immune cell apoptosis induction by trichostatin A and butyrate. Biochem Biophys Res Commun 247:833–837

Demonte D, Quivy V, Colette Y, Van Lint C (2004) Administration of HDAC inhibitors to reactivate HIV-1 expression in latent cellular reservoirs: implications for the development of therapeutic strategies. Biochem Pharmacol 68:1231–1238

Desilets A, Gheorghiu I, Yu SJ, Seidman EG, Asselin C (2000) Inhibition by deacetylase inhibitors of IL-1-dependent induction of haptoglobin involves CCAAT/Enhancer-binding protein isoforms in intestinal epithelial cells. Biochem Biophys Res Commun 276:673–679

Dinarello CA (1996) Biological basis for interleukin-1 in disease. Blood 87:2095–2147

Dinarello CA (1999) IL-18: a Th1-inducing, proinflammatory cytokine and new member of the IL-1 family. J Allergy Clin Immunol 103:11–24

Faggioni R, Jones-Carson J, Reed DA, Dinarello CA, Feingold KR, Grunfeld C, Fantuzzi G (2000) Leptin-deficient (ob/ob) mice are protected from T cell-mediated hepatotoxicity: role of tumor necrosis factor alpha and IL-18. Proc Natl Acad Sci U S A 97:2367–272

Fantuzzi G, Banda NK, Guthridge C, Vondracek A, Kim SH, Siegmund B, Azam T, Sennello JA, Dinarello CA, Arend WP (2003) Generation and characterization of mice transgenic for human IL-18-binding protein isoform a. J Leukoc Biol 74:889–896

Finnin MS, Donigian JR, Cohen A, Richon VM, Rifkind RA, Marks PA, Breslow R, Pavletich NP (1999) Structures of a histone deacetylase homologue bound to the TSA and SAHA inhibitors. Nature 401:188–193

Fusunyan RD, Quinn JJ, Fujimoto M, MacDermott RP, Sanderson IR (1999) Butyrate switches the pattern of chemokine secretion by intestinal epithelial cells through histone acetylation. Mol Med 5:631–640

Gibson PR, Rosella O, Wilson AJ, Mariadason JM, Rickard K, Byron K, Barkla DH (1999) Colonic epithelial cell activation and the paradoxical effects of butyrate. Carcinogenesis 20:539–544

Hsia SC, Shi YB (2002) Chromatin disruption and histone acetylation in regulation of the human immunodeficiency virus type 1 long terminal repeat by thyroid hormone receptor. Mol Cell Biol 22:4043–452

Huang N, Katz JP, Martin DR, Wu GD (1997) Inhibition of IL-8 gene expression in Caco-2 cells by compounds which induce histone hyperacetylation. Cytokine 9:27–36

Iwamoto M, Nara H, Hirata D, Minota S, Nishimoto N, Yoshizaki K (2002) Humanized monoclonal anti-interleukin-6 receptor antibody for treatment of intractable adult-onset Still's disease. Arthritis Rheum 46:3388–3389

van den Berg WB, Joosten LAB, Helsen M, Van de Loo FAJ (1994) Amelioration of established murine collagen-induced arthritis with anti-IL-1 treatment. Clin Exp Immunol 95:237–243

Van Lint C, Emiliani S, Ott M, Verdin E (1996) Transcriptional activation and chromatin remodeling of the HIV-1 promoter in response to histone acetylation. EMBO J 15:1112–1120

Vrana JA, Decker RH, Johnson CR, Wang Z, Jarvis WD, Richon VM, Ehinger M, Fisher PB, Grant S (1999) Induction of apoptosis in U937 human leukemia cells by suberoylanilide hydroxamic acid (SAHA) proceeds through pathways that are regulated by Bcl-2/Bcl-XL, c-Jun, and p21CIP1, but independent of p53. Oncogene 18:7016–7025

Webb Y, Zhou X, Ngo L, Cornish V, Stahl J, Erdjument-Bromage H, Tempst P, Rifkind RA, Marks PA, Breslow R, Richon VM (1999) Photoaffinity labeling and mass spectrometry identify ribosomal protein S3 as a potential target for hybrid polar cytodifferentiation agents. J Biol Chem 274:14280–14287

Wessells J, Baer M, Young HA, Claudio E, Brown K, Siebenlist U, Johnson PF (2004) BCL-3 and NF-kappaB p50 attenuate lipopolysaccharide-induced inflammatory responses in macrophages. J Biol Chem 279:49995–50003

Young JL, Sukhova GK, Foster D, Kisiel W, Libby P, Schonbeck U (2000) The serpin proteinase inhibitor 9 is an endogenous inhibitor of interleukin 1beta-converting enzyme (caspase-1) activity in human vascular smooth muscle cells. J Exp Med 191:1535–1544

4 Dendritic Cell Interactions and Cytokine Production

M. Foti, F. Granucci, P. Ricciardi-Castagnoli

4.1   Dendritic Cells Recognize Perturbations of the Immune System . . 63
4.2   Host–Pathogen Gene Profiling 66
4.3   Dendritic Cells and Pathogen Interaction 68
4.4   Dendritic Cells as Sensors of Infection 70
4.5   Dendritic Cell Transcriptional Profile Induced
        by Pathogens Include Cytokines and Chemokines 71
4.6   Activation of NK Cells by Dendritic Cells Is Mediated
        by Dendritic Cell-Derived IL-2 75
References 76

Abstract. The dendritic cell lineage comprises cells at various stages of functional maturation that are able to induce and regulate the immune response against antigens and thus function as initiators of protective immunity. The signals that determine the given dendritic cell functions depend mostly on the local microenvironment and on the interaction between dendritic cells and microorganisms. These interactions are complex and very different from one pathogen to another; nevertheless, both shared and unique responses have been observed using global genomic analyses. In this review, we have focused on the study of host–pathogen interactions using a genome-wide transcriptional approach with a focus on cytokine family members.

Immunity is the result of coevolution of microorganisms and the immune system; microorganisms have learned how to manipulate the immune response to their own advantage. Therefore, host–parasite interaction studies should reveal the molecular mechanisms that control the initiation, persistence, and polarization of immune responses.

In host–parasite interactions, dendritic cells (DCs) play a central role since they are located in close contact with the mucosal surfaces where they can sample incoming pathogens. If the amount of pathogen exceeds a certain threshold level for an extended period of time, DCs become activated and acquire a migratory capacity; during this "maturation" process, DCs undergo an extensive gene transcription reprogramming that involves the differential expression of up to 1,000 genes with the sequential acquisition of immune regulatory activities.

The immune response is extraordinarily complex, and it involves dynamic interaction of a wide array of tissues, cells and molecules. The diversity of innate immune mechanisms is in large part conserved in multicellular organisms (Mushegian and Medzhitov 2001). Some basic principles of microbial recognition and response are emerging, and recently, the application of computational genomics has played an important role in extending such observations from model organisms, such as Drosophila, to higher vertebrates, including humans. The analysis of gene expression in tissues, cells, and biological systems has evolved in the last decade from the analysis of a selected set of genes to an efficient high throughput whole-genome screening approach of potentially all genes expressed in a tissue or cell sample. Development of methodologies such as microarray technology allows an open-ended survey to identify comprehensively the fraction of genes that are differentially expressed between samples and define the samples' unique biology (Schena et al. 1995; Duggan et al. 1999; Lipshultz et al. 1999). This discovery-based research provides the opportunity to characterize either new genes with unknown function or genes not previously known to be involved in a biological process.

4.1 Dendritic Cells Recognize Perturbations of the Immune System

Dendritic cells are cells of the innate immunity characterized by broad functional properties (Fig. 1). These cells are divided into myeloid and plasmacytoid DC subsets, but the definition of dendritic cell subset phenotypes and the attribution of specific functions to defined DC stages has been a very difficult task; the definition of DC functional stages as mature versus immature or semi-mature DCs or activated versus steady-state DCs has raised a real language issue. In fact, DCs are characterized by a very high functional plasticity and can adapt their responses upon antigen encounter; they are also able to segregate in time different functions, which will dictate the outcome of the immune response.

DCs are located in nonlymphoid tissues, close to the mucosal surfaces, where they sample the environment to sense the infectious agents. For this task, DCs use a broad innate receptor repertoire and their phagocytic activity. These cells are named myeloid immature DCs and they are mostly resident in those tissues where they have originally seeded. Cells that have been conditioned by the microbial encounter migrate to

DCs regulate both innate and adaptive immunity by:

- ➢ Sampling the environment
- ➢ Activating inflammatory responses
- ➢ Activating NK cells
- ➢ Presenting self-and non self-antigens
- ➢ Activating appropriate T cell responses
- ➢ Regulating immune responses

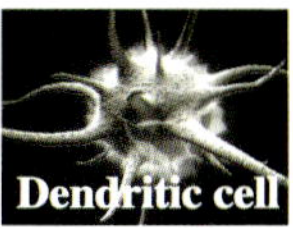

Fig. 1. DCs regulate both innate and adaptive immunity

the lymph nodes (LNs), where antigen is presented to specific T cells and where initiation of acquired immunity takes place. Thus, mature migratory DCs derive from the immature DCs, and are characterized by a limited degree of plasticity, a limited life span, and by the activation of an irreversible differentiation program ending with apoptotic cell death (Fig. 2).

The functional properties of immature resident DCs are characterized by the ability to sense the environment and to sample microbes at the mucosal sites. Indeed, DCs are particularly abundant in the respiratory tract and in the lungs as well as in the gut where, in addition to the Peyer's patches, they can be found in the lamina propria of the intestinal villi. We have shown that in order to sample the gut lumen and sense the intestinal flora, DCs are able to extend cytoplasmic protrusions in the lumen by opening the tight junctions of the epithelial cells and to preserve the integrity of the epithelial barrier by expressing, in a regulated way, tight

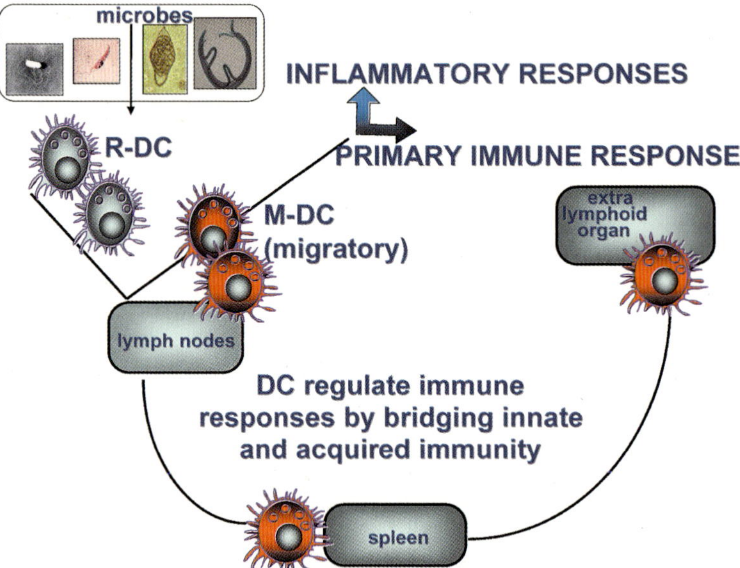

Fig. 2. DCs regulate immune responses by bridging innate and acquired immunity

junction proteins, such as the occludin (Rescigno et al. 2001). In the skin, the DCs, named Langerhans cells, form a tight network of cells continuously monitoring this tissue for invading parasites (Geissmann et al. 2002). Finally, in the liver and the spleen where blood-derived antigens are continuously brought and sampled, resident DCs are also particularly abundant.

To exert this sampling and sensing functional activities, DCs have a broad innate receptor repertoire for the recognition of infectious non-self antigens (Janeway and Medzhitov 2002). Indeed, in the mouse this receptor repertoire consists in the expression of the Toll-like receptor (TLR) family members that bind a variety of microbial ligands (Takeda et al. 2003). The signaling through these receptors is either mediated via the MyD88 adaptor molecule or via an independent pathway involving the TRIF molecule and the IRF-3 transcription factor (Beutler 2004). As a result of microbial activation through TLRs, the NFkB family members are activated and translocated to the nucleus (Hofer et al. 2001), inducing the transcription of many NFkB-dependent genes, mostly immune and inflammatory genes such as cytokines and chemokines. In addition, the microbial activation may also lead to the transcription of the interferon inducible genes via IRF-3 (Trottein et al. 2004). DCs can also display surface molecules such as the MARCO receptor that we have shown to be involved in a profound re-modeling of the actin cytoskeleton (Granucci et al. 2003b). Other receptors expressed by DCs are DC-SIGN and C-type lectin receptors capable of binding a variety of microorganisms, including viruses such as HIV and pathogenic bacteria such as *Mycobacterium tuberculosis* (Van Kooyk and Geijtenbeek 2003; Tailleux et al. 2003). Finally, resident DCs express the receptors for opsonized microbes such as FcR and CR, which have a key role in the phagocytosis process.

The study of host–pathogen interactions is instrumental for the control of infectious diseases. Host eukaryotes are constantly exposed to attacks by microbes seeking to colonize and propagate in host cells. To counteract them, host cells utilize a whole battery of defense systems to combat microbes. However, in turn, successful microbes evolved sophisticated systems to evade host defense. As such, interactions between hosts and pathogens are perceived as evolutionary arms races between genes of the respective organisms (Bergelson et al. 2001; Kahn et al. 2002; Woolhouse et al. 2002). Any interaction between a host and its pathogen involves alterations in cell signaling cascades in both partners. These alterations may be mediated by transcriptional or post-translational changes. The challenge of the postgenomic era is how to select target genes to be studied in detail from the thousands of genes encoded in the genome.

4.2   Host–Pathogen Gene Profiling

The interaction between a host and microbial pathogens are diverse and regulated. The molecular mechanisms of microbial pathogenesis show common themes that involve families of structurally and functionally related proteins such as adherence factors, secretion systems, toxins, and regulators of microbial pathogens. The interaction between pathogen and host uncovers unique mechanisms and molecules. Microarray expression analysis of pathogen infected cells and tissues can identify, simultaneously and in the same sample, host and pathogen genes that are regulated during the infectious process.

A major challenge to innate immune cells is the discrimination of foreign pathogens from self. As originally described by Janeway (Janeway 1989; Jandeway and Medzhitov 1998), innate immune cells such as DCs possess germline-encoded pattern recognition receptors (PRRs) that recognize and are triggered by evolutionary conserved molecules essential to pathogen function but absent in the host. These pathogen-associated molecular patterns (PAMPs) are widespread and include cell wall components such as mannans in the yeast cell wall, lipopolysaccharide (LPS) in Gram-negative bacteria, lipoproteins, peptidoglycans, and DNA containing unmethylated CpG motifs. There are at least two distinct classes of PRRs: those that mediate acute phagocytosis such as scavenger receptors and the mannose receptor and those that cause immediate cell activation such as TLRs. Upon cellular pathogen uptake, members of the TLR family become recruited to early phagosomes to screen their contents for ligands from foreign pathogens and subsequently to trigger cell activation upon ligand recognition (Underhill et al. 1999). This apparent "division of labor" (internalization versus cell activation) between scavenger receptors and

TLRs bears a caveat, given that cross-linking of the scavenger receptor CD36 profoundly modulates LPS-driven DC maturation (Urban et al. 2001).

Recent studies have shown a stereotyped range of host immune responses after infection with phylogenetically diverse organisms. Both bacterial and mammalian (mouse, human) genome sequences can be used in microarray technology to define the expression profile of pathogens and the host cells. The global transcription effects on host cells of the innate immunity by various bacterial pathogens, including *Listeria monocytogenes, Salmonella, Pseudomonas aeruginosa,* and *Bordetella pertussis* have been analyzed by using microarray technology (Rappuoli 2000). The infection of macrophages with *Salmonella typhimurium* identified novel genes whose level of expression are altered (Rosenberger et al. 2000). Similarly, *L. monocytogenes*-infected human promyelocytic THP1 cells identified 74 up-regulated RNAs and 23 down-regulated host RNAs (Cohen et al; 2000).

Many of the up-regulated genes encode proinflammatory cytokines (e.g., IL-8, IL-6), and many of the down-regulated genes encode transcriptional factors and cellular adhesion molecules. Understanding the molecular basis of the host response to bacterial infections is critical for preventing disease and tissue damage resulting from the host response. Furthermore, an understanding of host transcriptional changes induced by the microbes can be used to identify specific protein targets for drug development. PBMC transcriptomes have been studied by using cDNA microarrays after in-vitro stimulation with killed *Bordetella pertussis, Staphylococcus aureus,* and *Escherichia coli* (Boldrick et al. 2002). This study shows a core of 205 commonly expressed genes. These genes included those with both systemic and local effects. Highly represented were genes encoding intercellular immunoregulatory and signaling molecules such as cytokines and chemokines. These genes are regulated by NFκB, which orchestrates both innate and adaptive immune responses. Gram-negative bacteria induced stronger expression than Gram-positive bacteria. Moreover, the study describes 96 genes that were commonly repressed after a delay of about 2 h: a subset of monocyte-attracting chemokines, genes involved in cell–cell adhesion, diapedesis, and leucocyte extravasation, and those involved in recognizing bacteria and antigen presentation.

Therefore, it is possible to distinguish between different species and even individual strains of *B. pertussis* and *S. aureus*. The same group also showed that different expression responses to the same strain of *B. pertussis* depended on whether it was live or killed and, if live, whether it carried a toxin gene (Manger and Relman 2000). Also the host response to extracellular and intracellular parasites can be assessed by microarrays (Chaussabel et al. 2003; Blader et al. 2002; de Avalos et al. 2002). The gene expression program in response to *Trypanosoma cruzi* infection showed that while 106 genes were expressed at 24 h, none were induced more than twofold by 2 or 6 h. This was in contrast with a previous study investigating the host response to another intracellular pathogen, *Toxoplasma gondii*, where 63 known genes were up-regulated by 2 h after infection (Blader et al. 2001). The authors postulated that the time lag may be due to a parasite-dependent event that is required before the host cell "sees" and responds to the invasion with gene transcription, thus showing the use of microarrays in generating novel biological hypotheses.

Microarray technology can provide insights into the interaction between the pathogen and host by revealing global host expression responses to a range of pathogenic stimuli. Pathogens may manipulate host-cell gene expression, for example by causing upregulation of cellular support for pathogen replication and downregulation of MHC expression to allow pathogens to evade the immune system (de la Fuente et al. 2002; Shaheduzzaman et al. 2002). In conclusion, the amount of data generated by microarray experiments is enormous. Such quantities of data require statistical expertise and software to decipher patterns from these new expression repertoires, now stored in many different databases.

4.3 Dendritic Cells and Pathogen Interaction

When higher organisms are exposed to pathogenic microorganisms, innate immune responses occur immediately, both in terms of cell activation and inflammation. The initial response is characterized by uptake (that is, phagocytosis or endocytosis) and subsequent destruction or degradation of pathogens. At the initial stage of primary infection, DCs constitute an integral part of the innate immune system, supported by the activity of bone-marrow-derived nonspecific immune cells and various resident tissue cells. DCs and macrophages are acutely activated, and during this process they upregulate costimulatory cell surface molecules and major histocompatibility complex (MHC) class I and II molecules; they produce pro-inflammatory cytokines, such as tumor necrosis factor-α (TNF-α) and interleukins (IL1-β and effector cytokines, such as IL-12p40 and type I interferon); and they enhance presentation of the products of pathogen degradation (antigenic peptides) via the MHC class I or II presentation pathway to antigen reactive T cells (Aderem and Underhill 1999) and they produce bactericidal effector substances such as nitric oxide. Thus, innate immune cells and in particular DCs represent not only a first line of defense toward infections but also play an instructive role in shaping the adaptive immune responses (Fearon and Locksley 1996).

Adaptive immunity is controlled by the generation of MHC-restricted effector T cells and production of cytokines (Abbas et al. 1996). DCs are able to stimulate naïve T helper (Th) cells, which in turn may differentiate into Th1- versus Th2-polarized subsets; Th1 cells secrete primarily interferon IFNγ, whereas Th2 cells produce IL-4, IL-5, IL-10, and IL-13. Upon activation, DCs upregulate the expression of costimulatory molecules, such as CD80 and CD86, thereby increasing immunogenicity of peptide antigens presented. Finally, DC activation triggers production of cytokines, such as IL-12, IL-18, IL-4, or IL-10, that are able to polarize emerging T cell responses. Adding to the complexity, it is to date not clear whether all forms of activation of DCs necessarily result in increased immunogenicity. Furthermore, DCs can produce different cytokines in response to different activating stimuli (Moser and Murphy 2000). An example is shown by the observation that murine DCs phagocytosing either yeast or hyphae of *Candida albicans* produce either IL-12 or IL-4, and in vivo drive either Th1 or Th2 differentiation, respectively (d'Ostiani et al. 2000). Therefore DCs and macrophages are important at the interface in bridging the innate and adaptive immune system (Banchereau and Steinman 1998).

4.4 Dendritic Cells as Sensors of Infection

The immune system has developed mechanisms to detect and initiate responses to a continual barrage of immunological challenges. Dendritic cells play a major role as immunosurveillance agents. To accomplish this function, DCs are equipped with highly efficient mechanisms to detect pathogens, to capture, process, and present antigens, and to initiate T cell responses. The recognition of molecular signatures of potential pathogens, in DCs, is accomplished by membrane receptor of the toll-like family (TLRs), which activates dendritic cells, leading to the initiation of adaptive immunity. TLR signaling in DCs causes an increase in display of MHC peptide ligands for T cell recognition, upregulation of costimulatory molecules important for T cell clonal expansion and secretion of immunomodulatory cytokines, which direct T cell differentiation into effectors. Remarkably, ligation of distinct TLRs can trigger differential cytokine production in a single DC type or result in different cytokines in distinct DC subtypes. Studying the complexity of DC responses to TLR ligands illuminates the link between innate recognition and adaptive immunity, paving the way for improved vaccines and strategies to induce tolerance to autoantigens or allografts.

DCs comprise a distinct subset: human blood contains at least two distinct DC types, the myeloid DC CD11c+ and the plasmacytoid DC (PDC), as well as the monocyte precursors of Mon-DC (Shortman and Liu 2002). Unlike CD11c+ DC and Mon-DC, PDCs may have primarily an innate role in regulating antiviral responses, although they can also act as antigen-presenting cells. Mon-DC, monocytes, and neutrophils, expressed mRNA for TLRs 1, 2, 4, and 5 but only Mon-DCs expressed a TLR3 message (Muzio et al. 2000). Similarly, subsequent studies reported a decrease in expression of TLRs 1, 2, 4, 5, and 8 but an increase in TLR3 during monocyte differentiation into DCs (Visintin et al. 2001; Means et al. 2003). In contrast, human PDCs do not express a message for TLRs 2, 3, 4, 5, and 8 and are unresponsive to peptidoglycan, lipoteichoic acid, poly I:C, and LPS (Jarrossay et al. 2001; Kadowaki et al. 2001). PDCs express TLR9, which is not found in monocytes, granulocytes, Mon-DC, and CD11c+ DC.

The situation is different in mouse: mouse spleen PDCs express TLR7 and TLR9 but, in contrast to human, they also express mRNA for most other TLRs (Edwards et al. 2003). TLR9 is expressed by both murine plasmacytoid and nonplasmacytoid DCs. Thus, discussion of TLR repertoires should be restricted to particular cases where a given DC subset does not express detectable mRNA or respond to ligands for a particular TLR. We can use information about TLR repertoires and DC subset biology to predict some of the functions of TLRs in the immune system.

4.5   Dendritic Cell Transcriptional Profile Induced by Pathogens Include Cytokines and Chemokines

DCs have a crucial role in linking the class of immune response to the invading pathogen through the differential expression of T cell polarizing signals upon the ligation of selective pattern recognition receptors. The detailed mechanisms are still unknown. It is emerging that Th1 responses are initiated by intracellular TLR (TLR3, 7, 8, and 9) on DCs, resulting in high expression of the IL-12 gene family.

The microarray analysis has been used to study the DC transcriptome upon infection, by comparing the gene expression responses of dendritic cells to a bacterium (*E. coli*), a virus (influenza A), and a fungus (*C. albicans*) (Huang et al. 2001). In this work, a core of 166 genes that were induced by each organism in dendritic cells was described. The expression pattern of these genes indicated the sequence of events and coordination of pathways involved in immune responses.

Genes whose transcripts declined soon after pathogen contact include those involved in pathogen recognition and phagocytosis. Also at this stage, there was upregulation of genes expressing cytokines, chemokines, and immune cell receptors, which allows recruitment of other innate immune cells to the site of infection and genes modulating the cytoskeleton, which the authors postulated may be involved in dendritic cell migration. By 12 h after infection, there was increased expression of transcription factors and signaling molecules involved in lymphoid tissue regulation, antigen processing, and presentation. By 18 h, there was upregulation of chemokine receptor expression, thought to be related to the migration of dendritic cells to lymph nodes. In the time frame analyzed there was a sustained upregulation of production of reactive oxygen species, suggesting that there was continued killing of organisms by dendritic cells. This common core response was independent of pathogen characteristics and occurred in a coordinated fashion modulating innate and adaptive responses.

Expression analyses have shown that after microbial interaction, DCs undergo a multistep maturation process (Granucci et al. 2001) and acquire specific immune functions (Fig. 3), depending on the type of microbe they have encountered. The DC transcriptome has been described with a variety of different maturation stimuli. We have defined in detail the transcriptome induced in murine DCs by different pathogens such as *Schistosoma mansoni* and *Leishmania mexicana*. The data clearly demonstrate that individual parasites induce both common and individual regulatory networks within the cell. This suggests a mechanism whereby host-pathogen interaction is translated into an appropriate host inflammatory response.

Shistosoma mansoni, is a helminth parasite, has a complex life cycle that is initiated by the transcutaneous penetration of the larvae followed by its rapid transformation into schistosomula (SLA) (Pearce

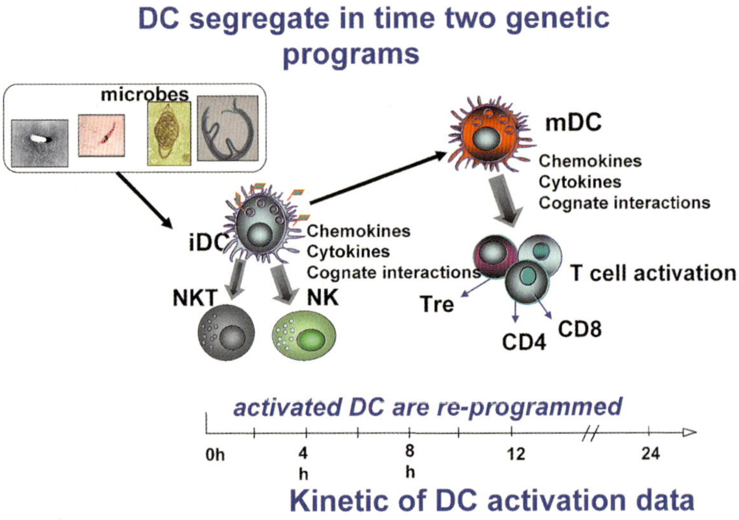

Fig. 3. DCs accomplish a number of tasks because they segregate in time two genetic programs

and MacDonald 2002; MadDonald et al. 2001). Once in the skin, SLA closely interact with immunocompetent cells, including DCs, to manipulate the host immune response (Ramaswamy et al. 2000; Angeli et al. 2001). SLA then begin a long vascular journey to reach the intrahepatic venous system, where they mature into adult male and egg-producing female worms. Eggs that accumulate in the liver, spleen, and lungs induce inflammation and an intense granulomatous hypersensitivity reaction (Rumbley and Phillips 1999). We have investigated DC–schistosome interactions using a genome-wide expression study. We have used a near-homogeneous source of mouse DCs, the well-defined, long term D1 splenic population (Winzler et al. 1997). The kinetic global gene expression analysis of mouse DCs stimulated with eggs or SLA indicated that genes encoding inflammatory cytokines, chemokines, and IFN-inducible proteins were oppositely regulated by the two stimuli (Trottein et al. 2004). Interestingly, eggs, but not SLA, induced the expression of IFN-β that efficiently triggered the type I IFN receptor (IFNAR) expressed on DCs, causing phosphorylation of STAT-1 with consequent upregulation of IFN-induced inflammatory products.

Clustering techniques applied to 283 differentially expressed genes revealed two signatures: the egg time-course experiment was compatible with a progressive cell differentiation process, such as maturation, whereas observations from SLA-stimulated DC samples suggested the occurrence of a stable blocking event within the first 4 h. Moreover, eggs modulated different amounts and subsets of genes in comparison with SLA, indicating that the two developmental stages of *S. mansoni* affected distinct intracellular pathways in DCs possibly by triggering specific receptors. The egg stage sustains the maximization of Ag presentation efficiency in DCs by inducing the upregulation of H-2M, which plays a crucial role in the peptide loading of MHC class II molecules (Kovats et al. 1998) and of the costimulatory molecules CD40 and ICAM-1. Cathepsins D and L, which are believed to remove the invariant chain from its complex with MHC class II molecules (Villadangos et al. 1999), are downregulated by SLA, but are not modulated by eggs, suggesting a reduction in the Ag processing capacity exerted by the larval stage on DCs. Moreover, the egg stage induced the expression of proinflammatory cytokine transcripts, such as TNF-α, and chemokines, such as

IP-10 (CXCL10), monocyte chemoattractant protein-5 (CCL12), MIP-1α (CCL3), MIP-1β (CCL4), MIP-1γ (CCL9), and MIP-2 (CXCL2), which are known to collectively attract granulocytes, immature DCs, NK cells, and activated T cells (Greaves and Schall 2000). *S. mansoni* eggs, but not SLA, induced the production of high amounts of IL-2, which could be important for DC-mediated activation of NK cells (Fernandez et al. 1999) or NKT cells (Fujii et al. 2002) as well as for priming naive T cells (Granucci et al. 2003a).

Mouse myeloid DCs, in response to helminth eggs, activate a strong interferon response compared to SLA. We have observed that the DC-derived IFNβ molecule efficiently triggered the IFNAR expressed on DCs, thus providing an autocrine and/or paracrine stimulation mechanism. Therefore, our data indicate myeloid DCs as one possible mediator of type I IFN signaling as well as one plausible source of IP-10 and MIP-1 production, also in response to helminth infections. The comparative gene expression analysis revealed two different DC global transcriptional modifications induced by either Schistosoma eggs or SLA, consistent with the different responses induced in vivo by these two parasite stages. Taken as a whole, these observations have provided new molecular insights into the host–parasite interaction established in the course of schistosomiasis leading to the identification of a type I IFN-dependent mechanism by which DCs may amplify inflammatory reactions in response to helminth infection.

As mentioned above, immature DCs express receptors that bind to microbes and microbial products, which are then internalized and processed for presentation to T cells. Whole bacteria and microbial products have all been found to induce cytokine and chemokines production by DCs. We have also used live bacteria to perturb immature DCs in the attempt to identify the kinetics of the DC reprogramming. DC functions are indeed determined according to a precise time frame. We observed that most of the changes in gene expression occur very early (2–4 h) after microbial encounter. A detailed analysis of the induced DC transcriptome has shown that at early time points following microbial exposure, the genes that are differentially expressed include those of the inflammatory and immune response pathway (IL1β, IL1ra, TNFα, IL-12 p40, IL-6, MIP-1α, β, γ, MIP-2α, I-309, C10, MCP-5, MIF, IP-10, GRO-1), as well as those involved in the interferon-inducible response.

In these transcriptional analyses, the most unexpected finding was the ability of R-DCs to produce IL-2 between 4 and 6 h after microbial exposure (Granucci et al. 2001). The IL-2 has been described as a NK, B, and T cell growth factor, and for this reason it may represent a key molecule conferring to DCs the unique ability to activate cells of the innate response. In agreement with this hypothesis, we have shown that IL-2-deficient DCs are severely impaired in their ability to activate NK cell responses both in vitro and in vivo as revealed by measuring antibacterial and anti-tumor responses (Aebischer et al. 2005).

4.6 Activation of NK Cells by Dendritic Cells Is Mediated by Dendritic Cell-Derived IL-2

The observation that DCs can produce IL-2 opens new possibilities for understanding the mechanisms by which DCs control innate immunity. Only microbial stimuli, and not inflammatory cytokines, are able to induce IL-2 secretion by DCs (Granucci et al. 2003a), indicating that this IL-2 production is tightly regulated by microbial signals and that DCs can distinguish between the actual presence of an infection and a cytokine-mediated inflammatory process. Thus, the adjuvant property of bacteria is explained by inducing in DCs immediate IL-2 production. This seems to be a unique feature of DCs, as macrophages are unable to produce IL-2 after bacterial activation. The ability of DCs to rapidly respond to microbial interaction with IL-2 production is also shown with parasites (e.g., *Leishmania mexicana*) and helminth (e.g., *Schistosoma* species) (Aebischer et al. 2005; Trottein et al. 2004). Of interest, only inflammatory stages of these two organisms (*Schistosoma* eggs or *Leishmania* promastigote) can induce IL-2 transcription in DCs (Aebischer et al. 2005; Trottein et al. 2004).

The cross-talk between DCs and NK cells has been described in the context of immune responses to infectious agents and tumors. IL-2 produced early by bacterially activated mouse DCs seems to play a fundamental role in the activation of NK cell-mediated immunity in vitro and in vivo (Granucci 2004). This indicates that in addition to its well-defined function in acquired immunity, IL-2 is also necessary, at least after bacterial infections, for the regulation of innate immune responses.

The biological relevance of NK cell activation mediated by DCs during bacterial infections resides mainly in the secretion of IFN-γ, which represents the principal phagocyte-activating factor. In addition, the secretion of IFN-γand TNF-α by activated NK cells in response to interaction with DCs provides a cytokine milieu promoting a strong cell-mediated innate immune response most beneficial in defense against microbial infections. Thus, DCs, through their ability to limit the spread of pathogens via innate cytokines, including IL-2 and type I IFNs, and to recruit innate cells via chemokine production, effectively participate in the early phases of the immune response. At later phases of the immune response, they also mediate immune-regulatory functions that will promote and shape acquired immune responses.

References

Abbas AK, Murphy KM, Sher A (1996) Functional diversity of helper T lymphocytes. Nature 383:787–793

Aderem A, Underhill DM (1999) Mechanisms of phagocytosis in macrophages. Annu Rev Immunol 17:593–623

Aebischer T, Bennett CL, Pelizzola M, Vizzardelli C, Pavelka N, Urbano M, Capozzoli M, Luchini A, Ilg T, Granucci F, Blackburn CC, Ricciardi-Castagnoli P (2005) A critical role for lipophosphoglycan in proinflammatory responses of dendritic cells to Leishmania mexicana. Eur J Immunol 35:476–486

Angeli V, FaveeuwC, Roye O, Fontaine J, Teissier E, Capron A, Wolowczuk I, Capron M, Trottein F (2001) Role of the parasite-derived prostaglandin D2 in the inhibition of epidermal Langerhans cell migration during schistosomiasis infection. J Exp Med 193:1135–1147

Bachem CW, van der Hoeven RS, de Bruijn SM, Vreugdenhil D, Zabeau M, Visser RG (1996) Visualization of differential gene expression using a novel method of RNA fingerprinting based on AFLP: analysis of gene expression during potato tuber development. Plant J 9:745–753

Banchereau J, Steinman RM (1998) Dendritic cells and the control of immunity. Nature 19:245–252

Bergelson J, Kreitman M, Stahl EA, Tian D (2001) Evolutionary dynamics of plant R-genes. Science 292:2281–2284

Beutler B (2004) Inferences, questions and possibilities in Toll-like receptor signaling. Nature 430:257–263

Birch PRJ, Kamoun S (2000) Studying interaction transcriptomes: coordinated analyses of gene expression during plant–microorganism interactions. In New technologies for life sciences: a trends guide (supplement to Elsevier Trends Journals, December 2000) London: Elsevier, pp 77–82

Blader IJ, Manger ID, Boothroyd JC (2001) Microarray analysis reveals previously unknown changes in Toxoplasma gondii-infected human cells. J Biol Chem 276:24223–24231

Boldrick JC, Alizadeh AA, Diehn M et al. (2002) Stereotyped and specific gene expression programs in human innate immune responses to bacteria. Proc Natl Acad Sci U S A 99:972–977

Brenner S, Johnson M, Bridgham J, Golda G, Lloyd DH, Johnson D et al. (2000) Gene expression analysis by massively parallel signature sequencing (MPSS) on microbead arrays. Nat Biotechnol 18:630–634

Chaussabel D, Tolouei SR, McDowell MA, Sacks D, Sher A, Nutman TB (2003) Unique gene expression profiles of human macrophages and dendritic cells to phylogenetically distinct parasites. Blood 102:672–681

Cohen P, Bouaboula M, Bellis M, Baron V, Jbilo O, Poinot-Chazel C, Galiegue S, Hadibi EH, Casellas P (2000) 190 Monitoring cellular responses to Listeria monocytogenes with oligonucleotide arrays. J Biol Chem 275:11181–11190

De Avalos SV, Blader IJ, Fisher M, Boothroyd JC, Burleigh BA (2002) Immediate/early response to Trypanosoma cruzi infection involves minimal modulation of host cell transcription. J Biol Chem 277:639–644

De la Fuente C, Santiago F, Deng L et al. (2002) Gene expression profile of HIV-1 Tat expressing cells: a close interplay between proliferative and differentiation signals. BMC Biochem 3:14

D'Ostiani CF, Del Sero G, Bacci A, Montagnoli C, Spreca A, E Mencacci A, Ricciardi-Castagnoli P, Romani L (2000) Dendritic cells discriminate between yeasts and hyphae of the fungus Candida albicans. Implications for initiation of T helper cell immunity in vitro and in vivo. J Exp Med 191:1661–1674

Duggan DJ, Bittner M, Chen Y, Meltzer P, Trent JM (1999) Expression profiling using cDNA microarrays. Nature Genet 21:10–14

Edwards AD, Diebold SS, Slack EMC, Tomizawa H, Hemmi H, Kaisho T et al. (2003) Toll-like receptor expression in murine DC subsets: lack of TLR7 expression by CD8$^+$ DC correlates with unresponsiveness to imidazoquinolines. Eur J Immunol 33:827–833

Fearon DT, Locksley RM (1996) The instructive role of innate immunity in the acquired immune response. Science 272:50–53

Fernandez NC, Lozier A, Flament C, Ricciardi-Castagnoli P, Bellet D, Suter M, Perricaudet M, Tursz T, Maraskovsky E, Zitvogel L (1999) Dendritic cells directly trigger NK cell functions: cross-talk relevant in innate anti-tumor immune responses in vivo. Nat Med 5:405–411

Fujii S, Shimizu K, Kronenberg M, Steinman RM (2002) Prolonged IFN-gamma-producing NKT response induced with alpha-galactosylceramide-loaded DCs. Nat Immunol 3:867–874

Geissmann F, Dieu-Nosjean MC, Dezutter C et al. (2002) Accumulation of immature Langerhans cells in human lymph nodes draining chronically inflamed skin. J Exp Med 196:417–430

Granucci F, C Vizzardelli N, Pavelka S, Feau M, Persico E, Virzi M, Rescigno G, Moro P, Ricciardi-Castagnoli (2001) Inducible IL-2 production by dendritic cells revealed by global gene expression analysis. Nat Immunol 2:882–888

Granucci F, Feau S, Angeli V, Trottein F, Ricciardi-Castagnoli P (2003a) Early IL-2 production by mouse dendritic cells is the result of microbial-induced priming. J Immunol 170:5075–5081

Granucci F, Petralia F, Urbano M et al. (2003b) The scavenger receptor MARCO mediates cytoskeleton rearrangements in dendritic cells and microglia. Blood. 102:2940–2947

Granucci F, Zanoni I, Pavelka N, Van Dommelen S, Andoniou C, Belardelli F, Degli Esposti MP, Ricciardi-Castagnoli P (2004) A contribution of mouse dendritic cell-derived IL-2 for NK cell activation. J Exp Med 200:287–295

Greaves DR, Schall TJ (2000) Chemokines and myeloid cell recruitment. Microb Infect 2:331–336

Hofer S, Rescigno M, Granucci F et al. (2001) Differential activation of NF-kappa B subunits in dendritic cells in response to Gram-negative bacteria and to lipopolysaccharide. Microbes Infect 3:259–265

Huang Q, Liu D, Majewski P et al. (2001) The plasticity of dendritic cell responses to pathogens and their components. Science 294:870–875

Janeway CA Jr (1989) Approaching the asymptote? Evolution and revolution in immunology. Cold Spring Harb Symp Quant Biol 54:1–13

Janeway CA Jr, Medzhitov R (1998) Introduction: the role of innate immunity in the adaptive immune response. Semin Immunol 10:349–350

Janeway CA Jr, Medzhitov R (2002) Innate immune recognition. Annu Rev Immunol 20:197–216

Jarrossay D, Napolitani G, Colonna M, Sallusto F, Lanzavecchia A (2001) Specialization and complementarity in microbial molecule recognition by human myeloid and plasmacytoid dendritic cells. Eur J Immunol 31:3388–3393

Kadowaki N, Antonenko S, Liu YJ (2001) Distinct CpG DNA and polyinosinic-polycytidylic acid double-stranded RNA, respectively, stimulate CD11c–type 2 dendritic cell precursors and CD11c+ dendritic cells to produce type I IFN. J Immunol 166:2291–2295

Kahn RA, Fu H, Roy CR (2002) Cellular hijacking: a common strategy for microbial infection. Trends Biochem Sci 27:308–314

Kamoun S, Hraber P, Sobral. B, Nuss D, Govers F (1999) Initial assessment of gene diversity for the oomycete pathogen Phytophthora infestans based on expressed sequences. Fungal Genet Biol 28:94–106

Kovats S, Grubin CE, Eastman S, deRoos P, Dongre A, Van Kaer L, Rudensky AY (1998) Invariant chain-independent function of H-2M in the formation of endogenous peptide-major histocompatibility complex class II complexes in vivo. J Exp Med 187:245–251

Liang P, Pardee AB (1992) Differential display of eukaryotic messenger RNA by means of the polymerase chain reaction. Science 257:967–971

Lipshultz RJ, Fodor SPA, Gingeras TR, Lockhart DJ (1999) High-density synthetic oligonucleotide arrays. Nature Genet 21:20–24

MacDonald AS, Araujo MI, Pearce EJ (2002) Immunology of parasitic helminth infections. Infect Immun 70:427–433

Manger ID, Relman DA (2000) How the host "sees" pathogens: global gene expression responses to infection. Curr Opin Immunol 12:215–218

Means TK, Hayashi F, Smith KD, Aderem A, Luster AD (2003) The Toll-like receptor 5 stimulus bacterial flagellin induces maturation and chemokine production in human dendritic cells. J Immunol 170:5165–5175

Moser M, Murphy KM (2000) Dendritic cell regulation of TH1-TH2 development. Nat Immunol 1:199–205

Mushegian A, Medzhitov R (2001) Evolutionary perspective on innate immune recognition. J Cell Biol. 155:705–710

Muzio M, Bosisio D, Polentarutti N, D'Amico O, Stoppacciaro A, Mancinelli R, et al. (2000) Differential expression and regulation of toll-like receptors (TLR) in human leukocytes: selective expression of TLR3 in dendritic cells. J Immunol 164:5998–6004

Pearce EJ, MacDonald AS (2002) The immunobiology of schistosomiasis. Nat Rev Immunol 2:499–511

Ramaswamy K, Kumar P, He YX (2000) A role for parasite-induced PGE2 in IL-10-mediated host immunoregulation by skin stage schistosomula of Schistosoma mansoni. J Immunol 165:4567–4574

Rappuoli R (2000) Pushing the limits of cellular microbiology: microarrays to study bacteria-host cell intimate contacts. Proc Natl Acad Sci U S A 97:13467–13469

Rescigno M, Urbano M, Valzasino B et al. (2001) Dendritic cells express tight junction proteins and penetrate gut epithelial monolayers to sample bacteria. Nat. Immunol 2:361–367

Rosenberger CM, Scott MG, Gold MR, Hancock RE, Finlay BB (2000) Salmonella typhimurium infection and lipopolysaccharide stimulation induce similar changes in macrophage gene expression. J Immunol 164:5894–5904

Rumbley CA, Phillips SM (1999) The schistosome granuloma: an immunoregulatory organelle. Microb Infect 1:499–504

Schena M, Shalon D, Davis RW, Brown PO (1995) Quantitative monitoring of gene expression patterns with a complementary DNA microarray. Science 270:467–470

Shaheduzzaman S, Krishnan V, Petrovic A et al. (2002) Effects of HIV-1 Nef on cellular gene expression profiles. J Biomed Sci 9:82–96

Shortman K, Liu Y (2002) Mouse and human dendritic cell subtypes. Nat Rev Immunol 2:151–161

Tailleux L, Schwartz O, Herrmann JL et al. (2003) DC-SIGN is the major Mycobacterium tuberculosis receptor on human dendritic cells. J Exp Med 197:121–127

Takeda K, Kaisho T, Akira S (2003) Toll like receptors. Annu Rev Immunol 21:335–376

Trottein F, Pavelka N, Vizzardelli C, Angeli V, Zouain C, Pellizzola M, Capron M, Belardelli F, Granucci F, Ricciardi-Castagnoli P (2004) A type I IFN-dependent pathway induced by Schistosoma mansoni eggs in mouse myeloid dendritic cells generates an inflammatory signature. J Immunol 172:3011–3017

Underhill DM, Ozinsky A, Hajjar AM, Stevens A, Wilson CB, Bassetti M, Aderem A (1999) The Toll-like receptor 2 is recruited to macrophage phagosomes and discriminates between pathogens. Nature 401:811–815

Urban BC, Willcox N, Roberts DJ (2001) A role for CD36 in the regulation of dendritic cell function. Proc Natl Acad Sci U S A 98:8750–8755

Van Kooyk Y, Geijtenbeek TB (2003) DC-SIGN: escape mechanism for pathogens. Nat Rev Immunol 3:697–709

Velculescu V, Zhang L, Zhou W, Vogelstein B, Basrai MA, Basette DE et al. (1997) Characterization of the yeast transcriptome. Cell 88:243–251

Villadangos JA, Bryant RA, Deussing J, Driessen C, Lennon-Dumenil AM, Riese RJ, Roth W, Saftig P, Shi GP, Chapman HA et al. (1999) Proteases involved in MHC class II antigen presentation. Immunol. Rev. 172:109–120

Visintin A, Mazzoni A, Spitzer JH, Wyllie DH, Dower SK, Segal DM (2001) Regulation of ToH-like receptors in human monocytes and dendritic cells. J Immunol 166:249–255

Winzler C, Rovere P, Rescigno M, Granucci F, Penna G, Adorini L, Zimmermann VS, Davoust J, Ricciardi-Castagnoli P (1997) Maturation stages of mouse dendritic cells in growth factor-dependent long-term cultures. J Exp Med 185:317–328

Woolhouse MEJ, Webster JP, Domingo E, Charlesworth B, Levin BR (2002) Biological and biomedical implications of the co-evolution of pathogens and their host. Nature Genet 32:569–577

5 Lysosomal Cysteine Proteases and Antigen Presentation

A. Rudensky, C. Beers

5.1   Introduction 82
5.2   The MHC Class II Pathway 82
5.3   Lysosomal Cysteine Proteases 85
5.4   MHC Class II Invariant Chain Degradation
       by Lysosomal Cysteine Proteases 87
5.5   Role of Cathepsin S and L in Generation
       of MHC Class II-Bound Peptides 89
5.6   Regulation of Cathepsin S and L Activity
       and Their Role in Ii Degradation in Macrophages 90
5.7   Role of Cathepsin S and L in MHC Class II Presentation
       by Nonprofessional Antigen-Presenting Cells 91
References 92

Abstract. Lysosomal proteinases are involved in two critical stages of MHC class II-mediated antigen presentation, i.e., degradation of invariant chain, a chaperone molecule critical for MHC class II assembly and transport, and generation of class II-binding peptides in the endocytic compartment. We found that two lysosomal cysteine proteinases, cathepsins S and L, were found to be differentially expressed in different types of "professional" and "nonprofessional" antigen presenting cells, including B cells, macrophages, specialized thymic epithelium, intestinal epithelium, and dendritic cells. In this chapter, our recent studies on the role of cathepsin S and L in MHC class II-mediated antigen processing and presentation in these cells types are highlighted.

5.1 Introduction

T lymphocytes recognize antigens in the form of short peptides bound to major histocompatibility complex (MHC) molecules displayed on the surface of antigen-presenting cells (APCs). MHC molecules sample products of proteolysis in two major sites of constitutive protein breakdown within the cells: MHC class I molecules present peptides generated in the cytosol and MHC class II molecules present peptides generated in the endosomal/lysosomal compartment. Peptides presented by MHC class I are recognized by $CD8^+$ T cells, while peptides generated in the endosomal/lysosomal compartment are presented in the context of MHC class II to $CD4^+$ T cells (Cresswell 1994).

$CD4^+$ T cell development is dependent upon recognition of self-MHC class II-peptide complexes in the thymus by TCR receptors expressed on immature thymocytes. Thymocytes capable of recognition of relatively low avidity MHC class II-self-peptide ligands expressed on cortical TECs proceed further with their differentiation program. Thymocytes displaying TCR with a high avidity for MHC class II-peptide complexes, expressed primarily by thymic bone marrow-derived cells, are subjected to deletion via apoptosis in the process dubbed negative selection. Elimination of these precursors of autoreactive T cells serves as a critical mechanism of autoimmunity prevention. Dendritic cells represent the most efficient type of bone marrow-derived antigen-presenting cells mediating negative selection in the thymus and presenting foreign and self-antigens in the periphery.

In this review, we discuss results of our recent studies of proteolytic mechanisms involved in MHC class II-mediated pathway of antigen presentation. Specifically, we focus on the role of lysosomal cysteine proteinases cathepsin S and L in maturation of MHC class II molecules and antigen processing in different types of antigen-presenting cells (APCs), including thymic cortical epithelial cells, dendritic cells, B cells, macrophages, and intestinal epithelial cells.

5.2 The MHC Class II Pathway

MHC class II molecules are constitutively expressed on the surface of professional antigen-presenting cells (APCs) of the immune system, i.e.,

B cells, dendritic cells (DCs), macrophages (Mφs), and cortical thymic epithelial cells (TECs), which present proteolytic fragments of self and foreign protein antigens to the $CD4^+$ T helper cells (Cresswell 1994; Watts 1997; Steimle et al. 1994). Expression of MHC class II can also be transiently induced by inflammatory signals in other cell types such as endothelial cells, fibroblasts, and different types of epithelial cells (Hershberg and Mayer 2000; Sartor 1994).

MHC class II molecules assemble in the endoplasmic reticulum with the help of the chaperone molecule invariant chain (Ii), a type II glycoprotein that promotes the proper folding and assembly of the MHC α/β heterodimer (Anderson and Miller 1992; Cresswell 1996) (Fig. 1). Alpha-beta heterodimers associate with the Ii binding to a particular region of Ii, termed CLIP (class II-associated Ii peptide), within the peptide-binding groove. The roles of Ii include stabilization of the nascent MHC class II heterodimers and prevention of the premature binding of polypeptides or partially folded proteins within the ER. Ii also facilitates intracellular trafficking of newly synthesized MHC class II molecules to the endosomes. The latter function of Ii is mediated by its cytoplasmic tail, which contains an endosomal/lysosomal targeting motif (Bakke and Dobberstein 1990; Lotteau et al. 1990; see for review Cresswell 1996). Upon arrival at the endosomal/lysosomal compartment, the invariant chain is degraded by lysosomal proteases in a step-wise fashion. Initial cleavage results in generation of the p22 and p18 Ii degradation intermediates followed by generation of the p12 fragment, all of which are still associated with the α/β heterodimer. At the late stages of Ii degradation, the p12 fragment is removed, leaving only CLIP associated with the MHC class II peptide binding groove (Maric et al. 1994; Neefjes and Ploegh 1992; Riese et al. 1996) (Fig. 2). The MHC class II-like chaperone molecule HLA-DM (H-2M in mice) mediates removal of CLIP in exchange for the diverse array of peptides derived from self or foreign protein antigens generated in the endosomes and lysosomes. Upon peptide binding, MHC class II-peptide complexes traffic to the plasma membrane via a poorly understood mechanism or retrograde transport from the late aspects of the endosomal/lysosomal compartment (Denzin and Cresswell 1995; Martin et al. 1996) (Fig. 1). Thus, lysosomal proteases are involved in two critical processes within

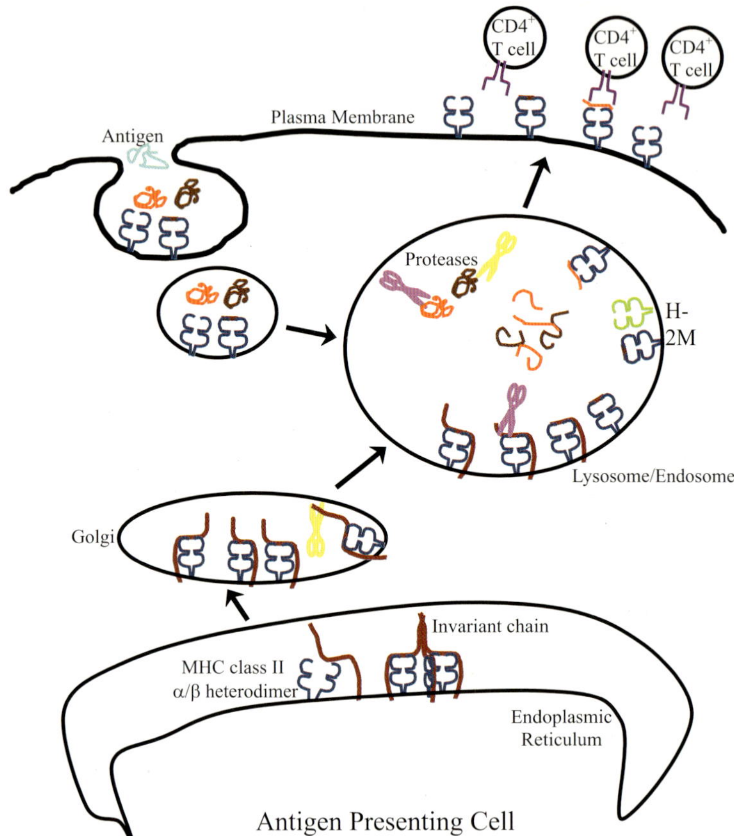

Fig. 1. The MHC class II presentation pathway. MHC class II α/β heterodimers assemble in the endoplasmic reticulum (ER) with the assistance of invariant chain (Ii). The cytoplasmic tail of Ii contains a motif that targets the MHC class II:Ii complex to the endosomal/lysosomal pathway. Maturation of the early endosomes leads to activation of lysosomal enzymes, including cysteinal proteases, which degrade endogenous and exogenous internalized proteins and Ii. Following Ii cleavage, the MHC class II peptide-binding groove remains occupied by the class II-associated invariant chain peptide (CLIP), which prevents peptide loading. Removal of CLIP and loading of peptides is mediated by the MHC-like molecule H-2M (DM). These newly assembled peptide:MHC class II complexes then traffic to the plasma membrane

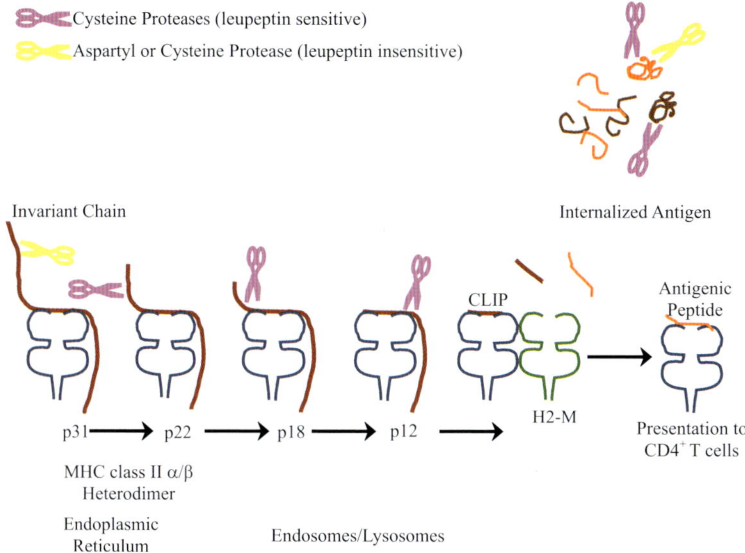

Fig. 2. Invariant chain degradation. Invariant chain (Ii) is degraded in a step-wise manner in the endosomes. The initial cleavage is likely mediated by a leupeptin-insensitive cysteine protease or by an aspartyl protease, whereas subsequent steps are mediated by leupeptin-sensitive cysteine proteases. Upon proteolytic removal of the bulk of Ii, MHC class II peptide-binding groove remains occupied by Ii-derived CLIP peptide (class II associated Ii peptide). In addition to Ii, lysosomal proteases degrade internalized exogenous and endogenous proteins in the endosomal/lysosomal compartment. MHC-like molecule H-2M (DM) facilitates exchange of CLIP for the peptides generated by these enzymes

MHC class II presentation pathway, i.e., degradation of invariant chain and generation of MHC class II binding peptides.

5.3   Lysosomal Cysteine Proteases

Lysosomal proteases belonging to a diverse protease family, known as the cathepsins, are involved in a number of important cellular processes. A number of these enzymes are cysteine proteases with an acidic pH optimum characteristic of the lysosomal compartment (Turk et al. 2000, 2001). Development of irreversible inhibitors binding to the active site cysteine residue allowed for utilization of their radio-labeled or biotinylated derivatives as specific probes to assess the levels of active cysteine proteases in different cell types, and even in different intracellular compartments (Bogyo et al. 2000; Driessen et al. 1999; Shi et al. 2000). Some lysosomal cysteine proteases, such as cathepsin B, are ubiquitously expressed in different tissues, whereas expression of others is restricted to certain tissue types (Chapman et al. 1997; Nakagawa and Rudensky 1999). Active cathepsin S (Cat S) can be detected in the thyroid and the spleen, but not in the thymic tissue, while active cathepsin L (Cat L) is found in kidney, liver, and thymic tissue but not in the spleen (Chapman et al. 1997; Nakagawa and Rudensky 1999; Villadangos et al. 1999). The distinct expression patterns of Cat S and Cat L activity in lymphoid tissues led us and others to further investigate the specific cell type distribution and to analyze the nonredundant function of these enzymes in MHC class II presentation pathway. In the course of these studies, we found Cat S activity in B cells and DCs, while Cat L activity was lacking in these cells. Cortical thymic epithelial cells revealed the opposite pattern of expression of these two enzymes upon investigation of mRNA, protein, and enzymatic activity using active site labeling. In contrast to the above cell types, Mφs express both Cat S and Cat L activity (Nakagawa et al. 1998, 1999; C. Beers and A. Rudensky, unpublished observations) (Fig. 3).

It has been anticipated that function of any given cathepsin is redundant, since none of the cathepsin-deficient mice studied so far exhibit a generalized defect in lysosomal protein degradation (Deussing et al. 1998; Nakagawa et al. 1998, 1999; Pham and Ley 1999; Roth et al. 2000; Saftig et al. 1998; Shi et al. 1999). However, our analysis of Cat L and Cat S knockouts as well as studies by others have demonstrated that these lysosomal cysteine proteases play a specific nonredundant function in the MHC class II pathway.

**Expression of Active Cathepsin S and Cathepsin L
in Professional APCs**

(detected by [125]labelled inhibitor active site labeling)

	Cathepsin S	Cathepsin L
Macrophage	+	+
B cells	+	−
Splenic/thymic DCs	+	−
cTECs	−	+

Fig. 3. Expression of Cathepsin S and Cathepsin L in professional antigen-presenting cells. Cysteine protease activity is detected using a substrate-analog inhibitor that covalently binds to the active site cysteine residue of the cathepsins. Macrophages, B cells, splenic and thymic DCs, and cTECs were purified from wild-type mice and incubated with the ^{125}I-labeled substrate-analog inhibitor for 3 h. Active cathepsin S and cathepsin L were detected by separating cellular lysates using 12% SDS/PAGE. ^{125}I-labeled cathepsins were visualized by autoradiography

5.4 MHC Class II Invariant Chain Degradation by Lysosomal Cysteine Proteases

Biochemical analyses of Cat S-deficient mice revealed a prominent accumulation of MHC class II-associated Ii degradation intermediates, p12 and p18, in Cat S-deficient B cells and DCs (Nakagawa et al. 1999; Shi et al. 1999). Proteolytic conversion of p12 Ii into CLIP is required for efficient loading of antigenic peptides and surface delivery of MHC class II peptide complexes. Therefore, antigen presentation studies utilizing a panel of MHC class II-restricted T cell hybridomas showed that the impaired Ii degradation in Cat $S^{-/-}$ bone marrow-derived APCs was associated with diminished presentation of the majority of exogenous antigens tested as compared to wild-type mice. Despite a defect in antigen presentation by B cells and DCs, the numbers of $CD4^+$ T cells remain normal in Cat $S^{-/-}$ mice (Nakagawa et al. 1998, 1999).

In contrast to Cat S-deficient mice, B cells and DCs from Cat L-deficient mice exhibit normal Ii degradation and MHC class II antigen processing and presentation consistent with the expression pattern of these two enzymes. However, a profound defect in Ii degradation became apparent upon analysis of Ii degradation intermediates in the thymic epithelium and purified cortical TECs, cells that mediate positive selection of CD4$^+$ T cells (Benavides et al. 2001; Nakagawa et al. 1998). These cells show impaired late-stage Ii degradation and a marked decrease in CD4$^+$ T cell numbers to about one-quarter of those present in littermate wild type animals. Overall expression of MHC class II on the cell surface of cortical TECs is normal. However, a substantial proportion of these MHC class II molecules is bound to p12 Ii fragments. Furthermore, the accumulation of Ii intermediates p12 and p18 within the endosomal/lysosomal compartment most likely prevents the proper loading of endogenous peptides, thereby limiting positive selection (Benavides et al. 2001; Nakagawa et al. 1998).

These data indicate that both Cat S and Cat L are required for late stages of Ii degradation in professional APCs, yet they function in distinct cell types. Cat L is critical for Ii processing in cortical TECs, cells that mediate positive selection of immature double-positive thymocytes, while Cat S plays a similarly critical role in bone marrow-derived cells that induce immune responses in the periphery and mediate central and peripheral tolerance. It is puzzling to speculate as to why there are two lysosomal cysteine proteases with an apparently similar role in the MHC class II presentation pathway, albeit functioning in separate cell types. One explanation is that these enzymes may also be involved in the generation of peptide fragments presented by MHC class II molecules. It can be further speculated that differences in the peptide repertoires displayed by positively selecting thymic cortical epithelial cells and tolerogenic dendritic cells in the thymus and in the periphery may be essential for development of a sizable peripheral CD4 T cell compartment and/or for diminishing potential risk of autoimmunity.

5.5 Role of Cathepsin S and L in Generation of MHC Class II-Bound Peptides

As discussed above, cathepsins have been implicated in the processing of exogenous and endogenous proteins for presentation by MHC class II molecules to CD4$^+$ T cells in addition to their role in late stages of Ii degradation.

We have taken a direct approach to determine the role of Cat S and Cat L in the generation of T cell epitopes by engineering fibroblast cell lines that express either Cat S or Cat L, or neither enzyme. Peptides were eluted from purified MHC class II molecules and analyzed using tandem mass spectrometry (Hsieh et al. 2002). It was observed that the majority of abundant peptides expressed by MHC class II molecules were not dependent upon expression of either Cat S or Cat L, although their relative expression varied in Cat S, Cat L, and control cells. However, a subset of MHC class II eluted peptides was significantly affected by the presence or absence of these enzymes. Specifically, the generation of some peptides required Cat S, while others were dependent on Cat L expression. The majority of peptides were generated in the presence of either Cat S or Cat L. In addition, the level of distinct subsets of MHC class II binding peptides was greatly diminished in the presence of Cat S or Cat L or both of these enzymes. In aggregate, studies utilizing these model APC lines indicate that Cat S and Cat L are able to influence endogenous MHC class II peptides.

By generating mice deficient in both Cat S or Cat L and Ii, the effect of these enzymes on generation of the MHC class II-binding peptide in vivo can be assessed directly as the lack of Ii abolishes the need to discriminate between the effects of these enzymes on the peptide processing from a defect in Ii degradation. The results of such an analysis indicate that Cat L influences the generation of T cell peptide epitope repertoire expressed by cortical TECs in vivo. Therefore, Cat L is essential not only for late stages of Ii degradation in thymic cortical epithelial cells, but also for generation of class II-bound peptides involved in positive selection of CD4 T cells. Cat S does not seem to have such a nonredundant role because overall peptide generation in DCs from mice lacking both Cat S and Ii appeared normal. However, generation of some peptide epitopes was affected. For example, Cat S was shown to destroy a known endogenously generated IgM T cell epitope (Honey et al. 2002; Hsieh et al. 2002). A separate study implicated Cat S in generating some, but not all hen egg lysozyme T cell epitopes in B cells (Pluger et al. 2002). In summary, these studies indicate that Cat L is involved in the generation of the majority of peptides involved in positive selection of $CD4^+$ T cells in cortical TECs, while Cat S influences some, but not the bulk, of MHC class II-bound peptides presented by DCs and B cells.

5.6 Regulation of Cathepsin S and L Activity and Their Role in Ii Degradation in Macrophages

Since macrophages display both Cat S and Cat L activity, we attempted to further investigate the relative involvement of these enzymes in MHC class II presentation in these cells, specifically, their contribution to the late stages of Ii degradation. In our earlier studies, we have observed a role for Cat S in regulating MHC II presentation of some exogenous antigens by Mφs (Nakagawa et al. 1999). In contrast, we failed to observe a detectable role for Cat L in invariant chain cleavage in IFN-γ-stimulated macrophages. In more recent experiments employing biosynthetic pulse-chase labeling of Cat S-, Cat L-deficient, or Cat S–Cat L double-deficient Mφs in combination with immunoprecipitation of MHC class II molecules, we confirmed that Cat S is the predominant enzyme processing Ii in these cells (Beers et al. 2003). Furthermore, we discovered that lack of a detectable role of Cat L in this process is due to a substantial downregulation of its enzymatic activity upon IFN-γ stimulation. Although this decline in Cat L activity coincided with a decrease in Cat L mRNA and an increase in secretion of mature Cat L protein, the level of intracellular mature Cat L protein did not change in IFN-γ stimulated pMφs. This finding suggests that Cat L activity is diminished due to induced expression of a Cat L-specific inhibitor. Previously, the p41 isoform of Ii was shown to exhibit Cat L-specific inhibitory activity due to the presence of a characteristic thyroglobulin-like domain. Since IFN-γ stimulation results in a drastic upregulation of MHC class II and associated accessory molecules including Ii, p41 was a logical candidate responsible for the observed inhibition of Cat L activity. However, we found that the p41 isoform of Ii did not contribute significantly to downregulation of Cat L activity in pMφs as comparable inhibition of Cat L activity was also observed in Ii-deficient pMφs upon IFN-γ induction. More recently, our preliminary studies implicate cystatin F as a specific inhibitor of Cat L activity in IFN-γ activated pMφs (C. Beers and A. Rudensky, unpublished observations). These results indicate that upon activation of pMφs in response to pro-inflammatory stimuli such as IFN-γ, enzymatic activity of Cat L is specifically inhibited, such that Cat S mediates Ii degradation and regulates MHC class II maturation. Notably, a similar phenomenon was observed in DCs isolated from mice expressing Cat L transgene driven by DC-specific CD11c promoter. In the latter cells, like in IFN-γ-stimulated pMφs, very little Cat L activity was observed, while high levels of the mature form of Cat L protein were present (Beers et al. 2003). Thus, during Th1 immune response dominated by IFN-γ production, inhibition of Cat L activity in macrophages results in preferential usage of cathepsin S for MHC class II presentation by all types of bone marrow derived antigen-presenting cells. This suggests that as a result of differential regulation of Cat L and Cat S, the latter enzyme governs MHC class II presentation by all bone marrow-derived APCs in secondary lymphoid organs.

These findings open up the possibility of manipulating the efficiency of antigen presentation by MHC class II molecules.

5.7 Role of Cathepsin S and L in MHC Class II Presentation by Nonprofessional Antigen-Presenting Cells

As mentioned above, MHC class II expression can also be induced by bacterial cell products and inflammatory cytokines on different types of nonprofessional APCs, including intestinal epithelium, fibroblasts, and endothelium (Hershberg and Mayer 2000; Sartor 1994). Interestingly, expression of MHC class II molecules on these nonprofessional APCs has been implicated in immune-mediated inflammation and progression of, or resistance to, autoimmunity (Hershberg and Blumberg 1999). In particular, it has been suggested that antigen presentation by intestinal epithelial cells (IECs) may play an important role in MHC class II-mediated presentation to immunoregulatory T cells (Campbell et al. 1999; Hershberg et al. 1997).

Epithelial cells at environmental interfaces provide protection from potentially harmful agents including pathogens. In addition to serving as a physical barrier and to producing soluble mediators of immunity such as antimicrobial peptides, these cells frequently function as "nonprofessional" antigen-presenting cells. In this regard, intestinal epithelial cells (IECs) are particularly prominent because they express MHC class II molecules at a site of massive antigenic exposure, where antigens could be presented to a large number of intraepithelial and lamina propria T lymphocytes expressing either $\gamma\delta$ or $\alpha\beta$ TCR. However, unlike bone marrow-derived professional APCs such as dendritic cells or B cells, little is known about the mechanisms of MHC class II presentation by IEC. Since thymic cortical epithelial cells, but not bone marrow-derived APCs, employ Cat L for Ii degradation and MHC class II peptide generation, we anticipated finding a similar role for Cat L in intestinal epithelial cells. Unexpectedly, we detected Cat S activity in epithelial cells of the small and large intestine. In contrast, Cat L activity was missing despite the presence of high levels of mature Cat L protein, indicating that Cat L activity is downmodulated. Further analysis revealed that Cat S is crucial for Ii degradation and antigen presentation by IEC (Beers et al. 2005).

In aggregate, our studies demonstrate that in vivo both professional and nonprofessional MHC class II-expressing APC utilize Cat S, but not Cat L, for MHC class II-mediated antigen presentation. In contrast, thymic cortical epithelial cells involved in positive selection of thymocytes are making use of Cat L for Ii degradation and generation of positively selecting ligands for CD4 T cells. The biological significance of differential expression of these two enzymes in distinct types of antigen-presenting cells in vivo remains unknown and needs to be addressed in future studies.

References

Anderson MS, Miller J (1992) Invariant chain can function as a chaperone protein for class II major histocompatibility complex molecules. Proc Natl Acad Sci U S A 89:2282–2286

Bakke O, Dobberstein B (1990) MHC class II-associated invariant chain contains a sorting signal for endosomal compartments. Cell 63:707–716

Beers C, Honey K, Fink S, Forbush K, Rudensky A (2003) Differential regulation of cathepsin S and cathepsin L in interferon gamma-treated macrophages. J Exp Med 197:169–179

Beers C, Burich A, Kleijmeer MJ, Griffith JM, Wong P, Rudensky A (2005) Cathepsin S controls MHC class II-mediated antigen presentation by epithelial cells in vivo. J Immunol 174:1205–1212

Benavides F, Venables A, Poetschke Klug H, Glasscock E, Rudensky A, Gomez M, Martin Palenzuela N, Guenet JL, Richie ER, Conti C J (2001) The CD4 T cell-deficient mouse mutation nackt (nkt) involves a deletion in the cathepsin L (CtsI) gene. Immunogenetics 53:233–242

Bogyo M, Verhelst S, Bellingard-Dubouchaud V, Toba S, Greenbaum D (2000) Selective targeting of lysosomal cysteine proteases with radiolabeled electrophilic substrate analogs. Chem Biol 7:27–38

Campbell N, Yio XY, So LP, Li Y, Mayer L (1999) The intestinal epithelial cell: processing and presentation of antigen to the mucosal immune system. Immunol Rev 172:315–324

Chapman HA, Riese RJ, Shi G P (1997) Emerging roles for cysteine proteases in human biology. Annu Rev Physiol 59:63–88

Cresswell P (1994) Assembly, transport, and function of MHC class II molecules. Annu Rev Immunol 12:259–293

Cresswell P (1996) Invariant chain structure and MHC class II function. Cell 84:505–507

Denzin LK, Cresswell P (1995) HLA-DM induces CLIP dissociation from MHC class II alpha beta dimers and facilitates peptide loading. Cell 82:155–165

Deussing J, Roth W, Saftig P, Peters C, Ploegh HL, Villadangos JA (1998) Cathepsins B and D are dispensable for major histocompatibility complex class II-mediated antigen presentation. Proc Natl Acad Sci U S A 95:4516–4521

Driessen C, Bryant RA, Lennon-Dumenil AM, Villadangos JA, Bryant PW, Shi GP, Chapman HA, Ploegh HL (1999) Cathepsin S controls the trafficking and maturation of MHC class II molecules in dendritic cells. J Cell Biol 147:775–790

Hershberg RM, Blumberg RS (1999) What's so (co)stimulating about the intestinal epithelium? Gastroenterology 117:726–728

Hershberg RM, Framson PE, Cho DH, Lee LY, Kovats S, Beitz J, Blum JS, Nepom GT (1997) Intestinal epithelial cells use two distinct pathways for HLA class II antigen processing. J Clin Invest 100:204–215

Hershberg RM, Mayer L F (2000) Antigen processing and presentation by intestinal epithelial cells – polarity and complexity. Immunol Today 21:123–128

Honey K, Nakagawa T, Peters C, Rudensky A (2002) Cathepsin L regulates CD4$^+$ T cell selection independently of its effect on invariant chain: a role in the generation of positively selecting peptide ligands. J Exp Med 195:1349–1358

Hsieh CS, deRoos P, Honey K, Beers C, Rudensky AY (2002) A role for cathepsin L and cathepsin S in peptide generation for MHC class II presentation. J Immunol 168:2618–2625

Lotteau V, Teyton L, Peleraux A, Nilsson T, Karlsson L, Schmid SL, Quaranta V, Peterson PA (1990) Intracellular transport of class II MHC molecules directed by invariant chain. Nature 348:600–605

Maric MA, Taylor MD, Blum JS (1994) Endosomal aspartic proteinases are required for invariant-chain processing. Proc Natl Acad Sci U S A 91:2171–2175

Martin WD, Hicks GG, Mendiratta SK, Leva HI, Ruley HE, Van Kaer L (1996) H-2M mutant mice are defective in the peptide loading of class II molecules, antigen presentation, and T cell repertoire selection. Cell 84:543–550

Nakagawa T, Roth W, Wong P, Nelson A, Farr A, Deussing J, Villadangos JA, Ploegh H, Peters C, Rudensky AY (1998) Cathepsin L: critical role in Ii degradation and CD4 T cell selection in the thymus. Science 280:450–453

Nakagawa TY, Brissette WH, Lira PD, Griffiths RJ, Petrushova N, Stock J, McNeish JD, Eastman SE, Howard ED, Clarke SR et al. (1999) Impaired invariant chain degradation and antigen presentation and diminished collagen-induced arthritis in cathepsin S null mice. Immunity 10:207–217

Nakagawa TY, Rudensky AY (1999) The role of lysosomal proteinases in MHC class II-mediated antigen processing and presentation. Immunol Rev 172:121–129

Neefjes JJ, Ploegh HL (1992) Intracellular transport of MHC class II molecules. Immunol Today 13:179–184

Pham CT, Ley T J (1999) Dipeptidyl peptidase I is required for the processing and activation of granzymes A and B in vivo. Proc Natl Acad Sci U S A 96:8627–8632

Pluger EB, Boes M, Alfonso C, Schroter CJ, Kalbacher H, Ploegh HL, Driessen C (2002) Specific role for cathepsin S in the generation of antigenic peptides in vivo. Eur J Immunol 32:467–476

Riese RJ, Wolf PR, Bromme D, Natkin LR, Villadangos JA, Ploegh HL, Chapman HA (1996) Essential role for cathepsin S in MHC class II-associated invariant chain processing and peptide loading. Immunity 4:357–366

Roth W, Deussing J, Botchkarev VA, Pauly-Evers M, Saftig P, Hafner A, Schmidt P, Schmahl W, Scherer J, Anton-Lamprecht I et al. (2000) Cathepsin L deficiency as molecular defect of furless: hyperproliferation of keratinocytes and perturbation of hair follicle cycling. FASEB J 14:2075–2086

Saftig P, Hunziker E, Wehmeyer O, Jones S, Boyde A, Rommerskirch W, Moritz JD, Schu P, von Figura K (1998) Impaired osteoclastic bone resorption leads to osteopetrosis in cathepsin-K-deficient mice. Proc Natl Acad Sci U S A 95:13453–13458

Sartor RB (1994) Cytokines in intestinal inflammation: pathophysiological and clinical considerations. Gastroenterology 106:533–539

Shi GP, Bryant RA, Riese R, Verhelst S, Driessen C, Li Z, Bromme D, Ploegh HL, Chapman HA (2000) Role for cathepsin F in invariant chain processing and major histocompatibility complex class II peptide loading by macrophages. J Exp Med 191:1177–1186

Shi GP, Villadangos JA, Dranoff G, Small C, Gu L, Haley KJ, Riese R, Ploegh HL, Chapman HA (1999) Cathepsin S required for normal MHC class II peptide loading and germinal center development. Immunity 10:197–206

Steimle V, Siegrist CA, Mottet A, Lisowska-Grospierre B, Mach B (1994) Regulation of MHC class II expression by interferon-gamma mediated by the transactivator gene CIITA Science 265:106–109

Turk B, Turk D, Turk V (2000) Lysosomal cysteine proteases: more than scavengers. Biochim Biophys Acta 1477:98–111

Turk V, Turk B, Turk D (2001) Lysosomal cysteine proteases: facts and opportunities. EMBO J 20:4629–4633

Villadangos JA, Bryant RA, Deussing J, Driessen C, Lennon-Dumenil AM, Riese RJ, Roth W, Saftig P, Shi GP, Chapman HA et al. (1999) Proteases involved in MHC class II antigen presentation. Immunol Rev 172:109–120

Watts C (1997) Capture and processing of exogenous antigens for presentation on MHC molecules. Annu Rev Immunol 15:821–850

6 DCs and Cytokines Cooperate for the Induction of Tregs

A.H. Enk

6.1 Interleukin-10 Modulates DCs for Tolerance Induction 98
6.2 IL-10-Dependent Feedback Mechanisms Between Treg and DCs . 99
6.3 TNFα and "Semi-mature" DCs 100
6.4 Suppressive Effects Mediated by TGF-β 101
6.5 Pharmaceuticals Interfere with DC Maturation 103
References 103

Abstract. Regulatory T cells (Treg) in broader terms consist of different subsets of T cells that are characterized by their ability to suppress proliferation of conventional effector T cells by various means. To date, three main groups of Treg can de distinguished, mainly by their functional properties (for review see Jonuleit and Schmitt 2003) Briefly, T regulatory (Tr)-1 cells as well as T helper (Th)-3 T cells express common T cell markers such as CD4 and are characterized by secretion of IL-10 and TGF-β, which provides a means by which proliferation of conventional CD4$^+$ cells is blocked. In contrast, genuine Treg that are characterized by their expression of CD25 block T cell proliferation by an unknown cell-to-cell contact-dependent mechanism. However, there are many overlapping features shared by the different subtypes of regulatory T cell and the common denominator is the production of regulatory cytokines such as IL-10 and TGF-β.

6.1 Interleukin-10 Modulates DCs for Tolerance Induction

IL-10 was originally described as cytokine-synthesis-inhibiting factor (CSIF) with regard to its effects exerted on IFNγ production of TH1 T cells. Meanwhile, it has been found to exert suppressive effects on a wide range of different populations of lymphocytes. When human or murine DCs are exposed to IL-10 in vitro culture systems, the cells display reduced surface expression of MHC class I and MHC class II molecules and reduced expression of T cell costimulatory molecules of the B7 family. In addition, the release of pro-inflammatory cytokines, i.e., IL-1α, IL-6, TNFα and most markedly IL-12, is abolished after IL-10 treatment (Griffen et al. 2001; ten Dijke and Hill 2004). However, all of these effects could only be recorded when immature DCs were exposed to IL-10. In contrast, mature DCs are insensitive to IL-10 and display a stable phenotype in the presence of IL-10 once they have matured (Gorelik et al. 2002; Skapenko et al. 2004).

According to their reduced MHC and B7 expression, the IL-10 treated DCs are inferior in T cell stimulation as opposed to their fully activated counterparts, but IL-10 does not merely keep DCs in an immature state; instead there is evidence that IL-10 modulates DC maturation, enabling DCs to induce T cells with regulatory properties. For example, freshly isolated Langerhans cells inhibit proliferation of TH1 cells after exposure to IL-10 but had no effect on TH2 cells (Szabo et al; 2002). Moreover, it has been shown that IL-10 modulated DCs from peripheral blood induce alloantigen specific anergy or anergy in melanoma-specific CD4$^+$ and CD8$^+$ T cells (Foussat et al; 2003; Cottrez et al; 2000). Further analysis of these anergic T cells revealed reduced IL-2 and IFN-γ production and in contrast to genuine Treg, reduced expression of the IL-2 receptor α-chain CD25. However, in addition to these anergic T cells, some authors have also observed the emergence of genuine Treg after injection of IL-10, as indicated by CD25$^+$ upregulation and cell-to-cell contact requirement for their suppressive activity (Wakkach et al. 2003).

The therapeutic use of these IL-10 modulated DCs is under investigation since injection of in vitro-generated, IL-10-modified DCs is able to prevent autoimmunity in a murine model of multiple sclerosis (EAE) and prolonged graft survival significantly in an murine GVHD model (Muller et al. 2002; Sato et al. 2003). Although most of these protocols involved in vitro exposure of DCs to IL-10, there is recent evidence that IL-10-driven DC modulation may also play a role in generation of regulatory T cells in vivo. For instance, Wakkach et al. were able to not only confirm previous in vitro results showing that addition of IL-10 to in vitro cultures differentiated DCs to a CD45^{high} tolerogenic phenotype, but they also demonstrated that this tolerogenic phenotype, along with regulatory Tr1 cells, is significantly enriched in spleens of IL-10 transgenic mice (Wakkach et al. 2003). Thus these data show that IL-10 plays an important role in rendering DCs not merely immature but also modifies their ability to induce regulatory T cells in vivo.

6.2   IL-10-Dependent Feedback Mechanisms Between Treg and DCs

Many results support the concept that DCs are inducers of Treg under certain circumstances. However, recent results imply that Treg, on the other hand, also affect DC functions (Serra et al. 2003). For example, Misra et al. have shown that DC cocultured with Treg remain in an immature state as judged by surface marker expression (Misra et al. 2004). These "Treg-exposed" DCs were inferior in induction of T cell proliferation and produced significant amounts of IL-10. In another murine cardiac transplantation model, increased numbers of splenic CD4$^+$/CD25$^+$ Treg and immature DCs were observed after treatment of the recipients with 15-deoxyspergualin, a commonly used antirejection drug (Min et al; 2003). As expected, these immature DCs purified from tolerant recipients induced the generation of CD4$^+$/CD25$^+$ Treg when incubated with naive T cells. Surprisingly, when these Treg isolated from tolerant recipients were incubated with DC progenitors, generation of DCs with tolerogenic properties, i.e., inferior T cell stimulatory capacity and IL-10 production, was observed. In conclusion, these results support the notion that IL-10 is a critical factor in a self-maintaining feed back loop, i.e., IL-10 derived from regulatory T cells has been shown to play a role in locking immature DCs in a tolerogenic state, which in turn induces further regulatory T cells that may contribute to IL-10 production (Misra et al. 2004). However, this positive feed back loop can ensure prolonged immunosuppression and does not only rely on the cell-to-cell contact required by genuine Treg.

6.3   TNFα and "Semi-mature" DCs

The term "immature" is not accurately defined in many aspects, and according to a long-standing definition, "real" immature DCs are only found in peripheral tissues; whereas the impetus to migrate toward regional lymph nodes requires at least some activation. Indeed there are reports showing that lung-derived "migratory" DCs (and hence "partly" activated DCs) account for the induction of regulatory T cells (Akbari et al. 2001). Therefore, tolerogenic DCs found in the lymph node may be differentially activated or "semi-mature".

In this regard, TNFα may play a role, since it has been shown that injection of DCs cultivated in presence of TNFα acted in a tolerogenic fashion (Menges et al. 2002). In these experiments, DCs were able to block autoimmunity in a murine model of multiple sclerosis (EAE). This suppressive effect was mediated by the induction of IL-10-producing regulatory T cells. The subsequent phenotypic analysis revealed that the DCs expressed regular amounts of MHC class II, and T cell costimulatory molecules, i.e., according to the authors, these DCs displayed a mature phenotype as judged by their surface marker expression. In contrast, these DCs failed to secrete IL1β, IL-6, TNFα, and in particular IL-12. The importance of IL-12 production for full maturation of DCs and acquisition of an immunstimulatory phenotype is further substantiated by results showing that IL-10 as well as cAMP are potent agonists of IL-12p70 secretion. And in fact, DCs treated with these agents are resistant to terminal maturation and induce T cell unresponsiveness in vitro (Griffin et al. 2001). In conclusion, maturity of DCs may not merely be judged by their surface marker expression, instead cytokine expression also has to be taken into account and only upregulation of several different indicators warrant a fully activated phenotype of DCs.

6.4   Suppressive Effects Mediated by TGF-β

Transforming growth factor (TGF)-β comprises different members of a larger family of similar molecules and can be divided into three closely related members: TGF-β1, TGF-β2, TGF-β3. All three members have an important role in regulation of an immune response. Most of the effects of these different TGFs are indistinguishable in in vitro cultures, because of their similar signaling pathway: i.e., all TGF-βs engage the same receptor complex that eventually activates transcription factors of the Smad family (ten Dijke and Hill 2004). Many studies have shown that TGF-β has the ability to block proliferation of effector T cells by limiting IL-2 production and upregulation of cell cycle inhibitors. Moreover, TGF-β can block differentiation of TH1 and TH2 subsets (Gorelik et al. 2002) by blocking IFN-β and IL-4 production, respectively. The latter is mediated by blocking the transcription factors T-bet (IFN-γ) and GATA-3 (IL-4) (Skapenko et al. 2004; Szabo et al. 2002).

These immunosuppressive qualities make TGF-β an ideal candidate molecule to convey suppression mediated by regulatory T cell. Despite its immunosuppressive activities, the direct involvement of TGF-β in effects induced by naturally occurring $CD25^+CD4^+$ Treg is still being debated since the main immunosuppressive mechanism seems to rely on cell-to-cell contact and is independent of soluble factors such as TGF-β.

However, so-called Tr1 cells display a unique cytokine profile since they neither express typical TH1 (IFN-β) nor TH2 (IL-4) cytokines, but produce sizable amounts of IL-10 and TGF-β. These cytokines are responsible for the suppressive effects exerted by Tr1 cells. Accordingly, via the secretion of TGF-β (and IL-10), Tr1 cells are able to suppress TH1- as well as TH2-mediated pathologies (Fousat et al. 2003; Cottrez et al; 2000). The differentiation of Tr1 cells is supposedly dependent on TGF-β too. It is thought that in a first step, naive $CD4^+$ T cells can be activated by antigen-presenting cells (APCs) in the presence of IL-10 and/or TGF-β, resulting in an anergic state that is already able to suppress effector T cells in a cell contact-dependent manner. However, in a second differentiation step, limited in vivo proliferation of these otherwise anergic T cells occurs and "fully differentiated" Tr1 cells will develop that produce IL-10 and TGF-β, but still remain unable to produce IL-2 and IL-4 (Levings et al. 2001).

Even more recently, a model has been proposed in which the production of TGF-β is a key factor leading to renewal of "Treg" in a self-sustained cascade of events. In this model, Zheng et al. (Zheng et al. 2004) showed that CD4$^+$/CD25$^-$ T cells develop into CD4$^+$/CD25$^+$ Treg after proliferation in presence of TGF-β. Thereafter these cells acquire immunosuppressive capacity along with the ability to secrete TGF-β and IL-10. As outlined above, these two cytokines are able to further generate "fresh" CD4$^+$/CD25$^+$ Tregs from a pool of naive CD4$^+$/CD25$^-$ T cells, which via secretion of TGF-β can then further propel the generation of third-, fourth-, fifth-, etc. generation Tregs.

Further evidence that TGF-β has potent effects on sustaining Treg-mediated suppression derives from the investigations of Jonuleit et al. (Jonuleit et al. 2002), which demonstrate the importance of TGF-β in so-called infectious tolerance. Here the authors have shown that CD4$^+$/CD25$^+$ Treg can "educate" naive CD4$^+$/CD25$^-$ T cells to acquire regulatory capabilities. Briefly, naive CD4$^+$ T cells were incubated with naturally occurring CD25$^+$ Treg, then recovered from these cocultures and finally coincubated with CD4$^+$ effector T cells. In stimulation assays, these precultured T cells were now able to suppress proliferation of the freshly isolated effector T cells. Thus, these data show that contact to naturally occurring CD25$^+$ Treg "infected" naive T cells to gain regulatory capacity. However, these second-generation regulatory T cells do not express the marker CD25 and the suppressive activity is cell-to-cell contact-independent. In fact, it has been shown that TGF-β is critically involved in mediating suppression of infectious tolerance. In summary, these data show that TGF-β plays a major role in maintenance of regulatory T cell function.

As outlined above, the effects of TGF-β on T cells is well documented; however, TGF-β also has immunosuppressive activity on APCs. To this end, it has been shown that Tr1-derived TGF-β is able to reduce the antigen presenting capacity of DCs in vitro (Geissmann et al. 1999) and Tr1 clones are able to limit production of immunoglobulin by B cells (Roes et al. 2003). Therefore, in addition to the direct effects on T cells, TGF-β can further augment immunosuppression in vivo by downregulating APC functions and thus further limiting the activation of effector T cells in vivo.

6.5 Pharmaceuticals Interfere with DC Maturation

In accordance with the concept that immature DCs induce Treg rather than effector T cells, several pharmaceuticals have been tested for their ability to induce Treg by affecting the maturation status of DCs. Among them are the vitamin D3 methobolite $1\alpha,25\text{-}(OH)_2D_3$, N-acethyl-L-cysteine and common immunosuppressive drugs such as corticosteroids, cyclosporin A, rapamycin, and aspirin (Greissman et al. 1999; Roes et al. 2003; Piemonti et al. 1999, 2000; Hackstein et al. 2001); Matyszak et al; 2000; Verhasselt et al. 1999). All of them have been shown to suppress DC maturation and as a consequence, anergy and/or regulatory T cells were induced. The effects are numerous and in the following examples are depicted only.

Direct induction of Treg in vitro by pharmacologically treated DCs has been observed after exposure of DCs to N-acetyl-L-cysteine and injection of DCs exposed to a mixture of vitamin D3 and mycophenolate mofetil, which induced full tolerance in an murine allograft model (Gregori et al. 2001). Interestingly, adoptive transfer of T cells from such tolerant mice into previously untreated mice prevented the rejection of respective allografts, thus indicating that probably regulatory T cells had been induced by vitamin D3-treated DCs in vivo. Furthermore, administration of rapamycin in clinically relevant doses prevented the full maturation of DCs and downregulated their IL-12 secretion and their capacity to induce T cell proliferation in vitro. Upon adoptive transfer of these rapamycin-treated DCs, an allo-antigen-specific T cell hyporesponsiveness could be observed in the recipients (Hackstein et al. 2003). In sum, there is abundant evidence showing that drugs affecting DC maturation by means of preventing DC maturation are also most likely inducers of Treg in vivo.

References

Akbari O, DeKruyff RH, Umetsu DT (2001) Pulmonary dendritic cells producing IL-10 mediate tolerance induced by respiratory exposure to antigen. Nat Immunol 2:725–731

Cottrez F, Hurst SD, Coffman RL, Groux H (2000) T regulatory cells 1 inhibit a Th2-specific response in vivo. J Immunol 165:4848–4853

Enk AH, Angeloni VL, Udey MC, Katz SI (1993) Inhibition of Langerhans cell antigen-presenting function by IL-A role for IL-10 in induction of tolerance. J Immunol 151:2390–2398

Foussat A, Cottrez F, Brun V, Fournier N, Breittmayer JP, Groux H (2003) A comparative study between T regulatory type 1 and CD4$^+$CD25$^+$ T cells in the control of inflammation. J Immunol 2003. 171:5018–5026

Geissmann F, Revy P, Regnault A, Lepelletier Y, Dy M, Brousse N, Amigorena S, Hermine O, Durandy A (1999) TGF-beta 1 prevents the noncognate maturation of human dendritic Langerhans cells. J Immunol 162:4567–4575

Gorelik L, Constant S, Flavell RA (2002) Mechanism of transforming growth factor beta-induced inhibition of T helper type 1 differentiation. J Exp Med 195:1499–1505

Gregori S, Casorati M, Amuchastegui S, Smiroldo S, Davalli AM, Adorini L (2001) Regulatory T cells induced by 1 alpha,25-dihydroxyvitamin D3 and mycophenolate mofetil treatment mediate transplantation tolerance. J Immunol 167:1945–1953

Griffin MD, Lutz W, Phan VA, Bachman LA, McKean DJ, Kumar R (2001) Dendritic cell modulation by 1alpha,25 dihydroxyvitamin D3 and its analogs: a vitamin D receptor-dependent pathway that promotes a persistent state of immaturity in vitro and in vivo. Proc Natl Acad Sci U S A 98:6800–6805

Hackstein H, Morelli AE, Larregina AT, Ganster RW, Papworth GD, Logar AJ, Watkins SC, Falo LD, Thomson AW (2001) Aspirin inhibits in vitro maturation and in vivo immunostimulatory function of murine myeloid dendritic cells. J Immunol 166:7053–7062

Hackstein H, Taner T, Zahorchak AF, Morelli AE, Logar AJ, Gessner A, Thomson AW (2003) Rapamycin inhibits IL-4-induced dendritic cell maturation in vitro and dendritic cell mobilization and function in vivo. Blood 101:4457–4463

Jonuleit H, Schmitt E, Kakirman H, Stassen M, Knop J, Enk AH (2002) Infectious tolerance: human CD25(+) regulatory T cells convey suppressor activity to conventional CD4(+) T helper cells. J Exp Med 196:255–260

Jonuleit H, Schmitt E (2003) The regulatory T cell family: distinct subsets and their interrelations. J Immunol 171:6323–6327

Levings MK, Sangregorio R, Roncarolo MG (2001) Human CD25(+)CD4(+) T regulatory cells suppress naive and memory T cell proliferation and can be expanded in vitro without loss of function. J Exp Med 193:1295–1302

Matyszak MK, Citterio S, Rescigno M, Ricciardi-Castagnoli P (2000) Differential effects of corticosteroids during different stages of dendritic cell maturation. Eur J Immunol 30:1233–1242

Menges M, Rossner S, Voigtlander C, Schindler H, Kukutsch NA, Bogdan C, Erb K, Schuler G, Lutz MB (2002) Repetitive injections of dendritic cells matured with tumor necrosis factor alpha induce antigen-specific protection of mice from autoimmunity. J Exp Med 195:15–21

Min WP, Zhou D, Ichim TE, Strejan GH, Xia X, Yang J, Huang X, Garcia B, White D, Dutartre P, Jevnikar AM, Zhong R (2003) Inhibitory feedback loop between tolerogenic dendritic cells and regulatory T cells in transplant tolerance. J Immunol 170:1304–1312

Misra N, Bayry J, Lacroix-Desmazes S, Kazatchkine MD, Kaveri SV (2004) Cutting edge: human CD4$^+$CD25$^+$ T cells restrain the maturation and antigen-presenting function of dendritic cells. J Immunol 172:4676–4680

Muller G, Muller A, Tuting T, Steinbrink K, Saloga J, Szalma C, Knop J, Enk AH (2002) Interleukin-10-treated dendritic cells modulate immune responses of naive and sensitized T cells in vivo. J Invest Dermatol 119:836–841

Piemonti L, Monti P, Allavena P, Sironi M, Soldini L, Leone BE, Socci C, Di C, V (1999) Glucocorticoids affect human dendritic cell differentiation and maturation. J Immunol 162:6473–6481

Piemonti L, Monti P, Sironi M, Fraticelli P, Leone BE, Dal Cin E, Allavena P, Di Carlo V (2000) Vitamin D3 affects differentiation, maturation, and function of human monocyte-derived dendritic cells. J Immunol 164:4443–4451

Roes J, Choi BK, Cazac BB (2003) Redirection of B cell responsiveness by transforming growth factor beta receptor. Proc Natl Acad Sci U S A 100:7241–7246

Sato K, Yamashita N, Matsuyama T (2002) Human peripheral blood monocyte-derived interleukin-10-induced semi-mature dendritic cells induce anergic CD4(+) and CD8(+) T cells via presentation of the internalized soluble antigen and cross-presentation of the phagocytosed necrotic cellular fragments. Cell Immunol 215:186–194

Sato K, Yamashita N, Baba M, Matsuyama T (2003) Modified myeloid dendritic cells act as regulatory dendritic cells to induce anergic and regulatory T cells. Blood 101:3581–3589

Serra P, Amrani A, Yamanouchi J, Han B, Thiessen S, Utsugi T, Verdaguer J, Santamaria P (2003) CD40 ligation releases immature dendritic cells from the control of regulatory CD4$^+$CD25$^+$ T cells. Immunity 19:877–889

Skapenko A, Leipe J, Niesner U, Devriendt K, Beetz R, Radbruch A, Kalden JR, Lipsky PE, Schulze-Koops H (2004) GATA-3 in human T cell helper type 2 development. J Exp Med 199:423–428

Szabo SJ, Sullivan BM, Stemmann C, Satoskar AR, Sleckman BP, Glimcher LH (2002) Distinct effects of T-bet in TH1 lineage commitment and IFN-gamma production in CD4 and CD8 T cells. Science 295:338–342

Ten Dijke P, Hill CS (2004) New insights into TGF-beta-Smad signalling. Trends Biochem Sci 29:265–273

Verhasselt V, Vanden Berghe W, Vanderheyde N, Willems F, Haegeman G, Goldman M (1999) N-acetyl-L-cysteine inhibits primary human T cell responses at the dendritic cell level: association with NF-kappaB inhibition. J Immunol 162:2569–2574

Wakkach A, Fournier N, Brun V, Breittmayer JP, Cottrez F, Groux H (2003) Characterization of dendritic cells that induce tolerance and T regulatory 1 cell differentiation in vivo. Immunity 18:605–617

Zheng SG, Wang JH, Gray JD, Soucier H, Horwitz DA (2004) Natural and induced $CD4^+CD25^+$ cells educate $CD4^+CD25^-$ cells to develop suppressive activity: the role of IL-2, TGF-beta, IL-10. J Immunol 172:5213–5221

7 Induction of Immunity and Inflammation by Interleukin-12 Family Members

G. Alber, S. Al-Robaiy, M. Kleinschek, J. Knauer, P. Krumbholz,
J. Richter, S. Schoeneberger, N. Schuetze, S. Schulz, K. Toepfer,
R. Voigtlaender, J. Lehmann, U. Mueller

7.1 Structure of IL-12 Family Members 108
7.2 Expression of IL-12 Family Members and Their Receptors 109
7.3 Signal Transduction 111
7.4 Biology of IL-12 Family Members 112
7.5 Role of IL-12 Family Members in Host Defense 114
7.6 Antitumor Activity of IL-12 Family Members 116
7.7 Role of IL-12 Family Members in Organ-Specific Autoimmunity . 117
7.7.1 Experimental Autoimmune Encephalomyelitis 118
7.7.2 Rheumatoid Arthritis 118
7.7.3 Psoriasis Vulgaris 119
7.7.4 Inflammatory Bowel Disease 120
7.7.5 Insulin-Dependent Diabetes Mellitus 121
References 121

Abstract. The interleukin (IL)-12 family is composed of three heterodimeric cytokines, IL-12 (p40p35), IL-23 (p40p19), and IL-27 (EBI3p28), and of monomeric and homodimeric p40. This review focuses on the three heterodimeric members of the IL-12 family. The p40 and p40-like (EBI3) subunits have homology to the IL-6R, the other subunits (p35, p19, and p28) are homologous to each other and to members of the IL-6 superfamily. On the basis of their structural similarity, it was expected that the members of the IL-12 family have overlapping pro-inflammatory and immunoregulatory functions. However, it was surprising that they also show very distinct activities. IL-12 has a central role as a Th1-inducing and -maintaining cytokine, which is essential in cell-mediated immunity in nonviral infections and in tumor control. IL-23 recently emerged as an end-stage effector cytokine responsible for autoimmune chronic inflammation through induction of IL-17 and direct activation of macrophages. Very recently, IL-27 was found to exert not only a pro-inflammatory Th1-enhancing but also a significant anti-inflammatory function.

7.1 Structure of IL-12 Family Members

Interleukin (IL)-12 was discovered as "natural killer cell stimulatory factor (NKSF)" in 1989 (Kobayashi et al. 1989) and by another group as "cytotoxic lymphocyte maturation factor (CLMF)" in 1990 (Stern et al. 1990). It was the first cytokine to be found with a heterodimeric structure (Gately et al. 1998; Trinchieri 1998). IL-12 is composed of a p35 and a p40 subunit. Both subunits are covalently linked. Each subunit is expressed by its own gene, whereby the *p35* and *p40* genes are located on different chromosomes. The sequence of p35 shows homology with that of IL-6 and G-CSF (Merberg et al. 1992) and encodes a p35 subunit that shows a four-α-helix bundle structure typical of cytokines. The sequence of the p40 subunit is homologous to the extracellular portion of members of the hemopoietin receptor family, in particular to the IL-6 receptor α-chain (IL-6Rα) and the ciliary neurotrophic factor receptor (CNTFR) (Gearing and Cosman 1991). Therefore, IL-12 represents a heterodimeric cytokine with one chain (p35) having the actual cytokine structure and the other chain (p40) showing the structure of a cytokine receptor. While IL-12 might have evolved from the IL-6/IL-6R family, it has meanwhile established its own family (Fig. 1). In 2000, a novel p40p19 heterodimeric molecule was found and designated IL-23 (Oppmann et al. 2000). Only 2 years later, another IL-12 family member termed IL-27 was described, which consists of the p40-like subunit EBI3 noncovalently linked to a p28 subunit (Pflanz et al. 2002). Thus, the novel IL-12 family consists of cytokines with a p40 or a p40-like (EBI3) receptor component together with members of the long-chain four-α-helix bundle cytokine family (p35, p19, and p28) (Fig. 1). This

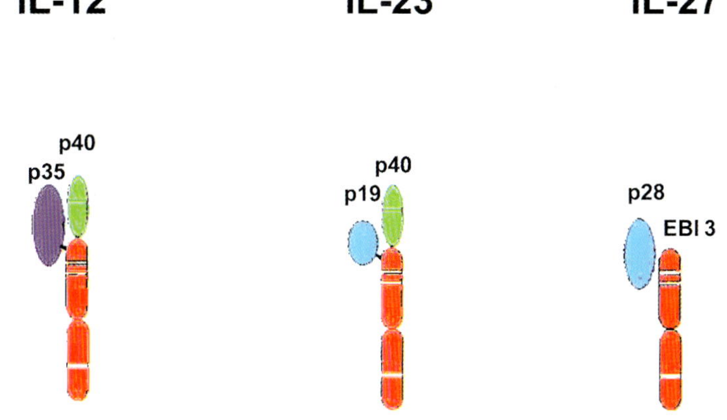

Fig. 1. Members of the heterodimeric IL-12 family. This figure focuses on the three heterodimeric IL-12 family members and does not show monomeric and homodimeric IL-12p40, which can be secreted in molar excess relative to the secretion of IL-12 and IL-23. The four-α-helix bundle cytokine p35 is shown in *purple*, p19 and p28 are shown in *blue*. For p40 and EBI3, an Ig-like domain (*green*) and cytokine receptor homology domains (*red*) are shown. Parts of Figs. 1–3 have kindly been provided by Rob Kastelein (DNAX Research Institute)

review focuses on the three heterodimeric IL-12 family members and does not elaborate on monomeric and homodimeric IL-12p40 with their antagonistic and agonistic properties.

7.2 Expression of IL-12 Family Members and Their Receptors

All IL-12 family members are produced by activated dendritic cells (DCs) and macrophages, which function as antigen-presenting cells. Expression of p35, p19, and p28 is found in many different cell types. However, p40 transcription appears to be restricted to antigen-presenting cells. Coexpression of both chains of IL-12, IL-23, and IL-27 in one cell is required to generate the bioactive form of either member of the IL-12 family (Gubler et al. 1991; Oppmann et al. 2000; Pflanz et al.

2002). Free subunits of p35, p19, and p28 chains are not or only inefficiently secreted in humans or mice. In contrast, p40 can be secreted as monomer, homodimer, or polymer in a 10- to 1,000-fold excess relative to heterodimeric IL-12 (D'Andrea et al. 1992). Stimuli for expression of IL-12 family members include pathogen-associated molecular patterns (PAMPs), which are ligands for toll-like receptors (TLRs) on phagocytes and DCs. In addition, optimal production of IL-12 (and probably also of IL-23 and IL-27) requires cytokines (IFN-γ, IL-4, IL-13) (Trinchieri 2003). Activated T cells can enhance IL-12 production not only by cytokine secretion but also by direct cell-to-cell contact via CD40L/CD40 interaction (Schulz et al. 2000).

Target cells for all IL-12 family members are natural killer (NK) and T cells. In addition, macrophages and DCs appear to express functional receptors for IL-12 and IL-23 (Belladonna et al. 2002; Grohmann et al. 1998, 2001). For IL-27, a receptor has also been described on monocytes, Langerhans' cells, activated DCs, and endothelial cells (Pflanz et al. 2004; Villarino et al. 2004). The activities of IL-12 are mediated by a high-affinity (50-pM) receptor composed of two subunits, designated IL-12Rβ1 and IL-12Rβ2 (Gately et al. 1998) (Fig. 2). Both receptor chains are members of the class I cytokine receptor family and are most closely related to glycoprotein gp130, the common receptor β-chain of the IL-6-like cytokine superfamily. IL-23 was found to bind to IL-12Rβ1 but not IL-12Rβ2 (Oppmann et al. 2000). Instead, a novel gp130-like subunit designated IL-23R was identified, which together with IL-12Rβ1 allows for high-affinity binding of IL-23 (Parham et al. 2002). One chain of the IL-27 receptor complex is the WSX-1/TCCR chain (Pflanz et al. 2002). In 2004, the second chain of the IL-27 receptor complex was found to be gp130 (Pflanz et al. 2004).

The individual members of the IL-12 family have overlapping, but also distinct, activities. This may partially be based on different receptor components expressed on different target cells or during different developmental stages of the target cells (naïve vs. memory Th cells).

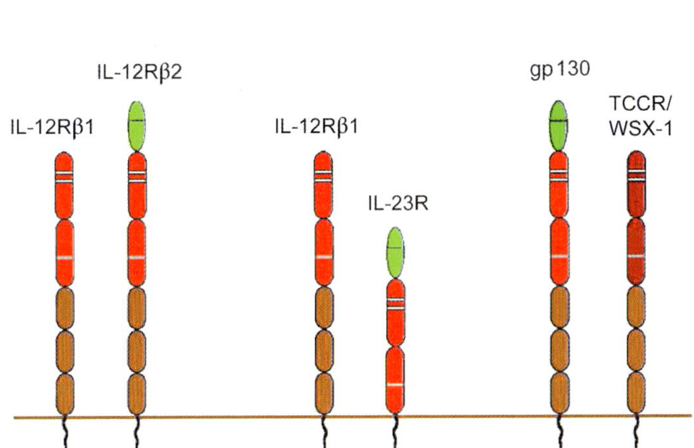

Fig. 2. Receptors for the members of the IL-12 family. The receptor subunits can carry an Ig-like domain (*green*), cytokine receptor homology domains (*red*) and fibronectin-like domains (*brown*)

7.3 Signal Transduction

IL-12 induces via its IL-12Rβ1/β2 receptor complex tyrosine phosphorylation of the Janus-family kinases Jak2 and Tyk 2, which in turn activate signal transducer and activator of transcription (STAT) 1, STAT3, STAT4, and STAT5 (Presky et al. 1996) (Fig. 3). Activation of STAT4 is essential for the specific cellular effects of IL-12. STAT4$^{-/-}$ mice show a phenotype that is identical to the phenotype of IL-12p35$^{-/-}$ mice (Kaplan et al. 1996).

Binding of IL-23 to its receptor complex IL-12Rβ1/IL-23R results in tyrosine phosphorylation of Jak2 and Tyk2, leading to activation of STAT1, STAT3, and STAT4. In contrast to the IL-12-induced predominant formation of STAT4 homodimers, IL-23 rather induces STAT3/STAT4 heterodimers (Fig. 3) (Parham et al. 2002). Formation of different DNA-binding STAT dimers by IL-12 and IL-23 appears to contribute to their distinct functional properties.

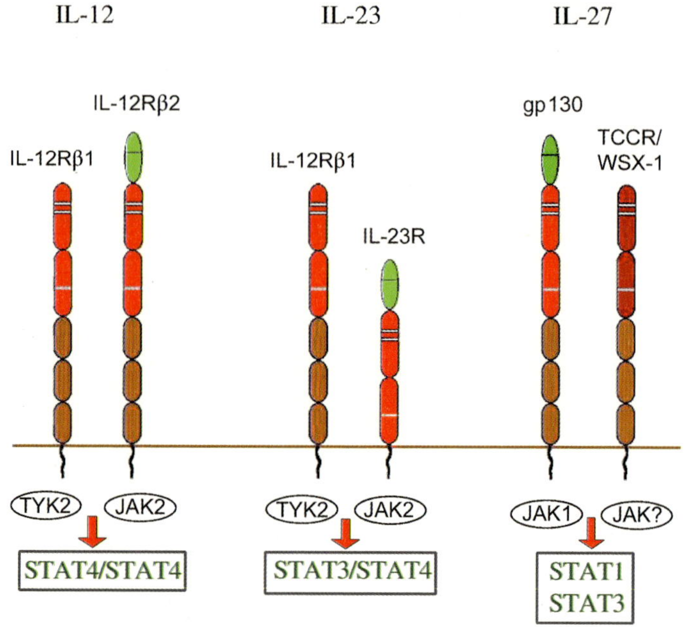

Fig. 3. Signal transduction components activated by IL-12 family members. Differential formation of STAT dimers might be essential for the distinct activities of the individual members of the IL-12 family

Upon binding of IL-27 to its receptor complex WSX-1/gp130, Jak1, STAT1 and STAT3 are activated (Pflanz et al. 2004). Activation of STAT1 and STAT3 by IL-27 might be the basis for its capability of downregulating T cell hyperactivity and IFN-γ production (Villarino et al. 2003; Hamano et al. 2003). In addition, it is of particular interest that IL-27-induced activation of STAT1 can directly induce expression of the transcription factor T-bet, which can regulate IL-12 responsiveness of T cells by upregulating IL-12Rβ2 (Takeda et al. 2003). Current and future research will help to further clarify the pleiotropic effects of IL-27-dependent signaling and link it to biological functions.

7.4 Biology of IL-12 Family Members

The biological functions of the individual members of the IL-12 family are surprisingly distinct and overlap only partially. Initially after the discovery of the latest members of this family, very similar immunological functions were expected of the individual family members based on their structural resemblance. However, more recent studies indicate distinct activities. The three heterodimeric members of the IL-12 family are pro-inflammatory cytokines. IL-12 represents the prototypic molecule for induction and maintenance of T helper 1 (Th1) cells secreting IFN-γ. On the basis of the expression of the individual receptor complexes for IL-12, IL-23, and IL-27, it appears that the three family members act on Th cells *at different stages of the Th1 differentiation pathway.* Naïve Th cells express the receptors for IL-27 and IL-12, but not for IL-23 (Brombacher et al. 2003). On the other hand, memory Th cells express the receptor for IL-23, but have only low or no expression of the receptors for IL-27 or IL-12 (Oppmann et al. 2000; Chen et al. 2000). Mullen et al. showed that T-bet expression in committed Th1 cells precedes exposure to IL-12 (Mullen et al. 2001). This argues for a role of IL-27 for the induction and IL-12 for the maintenance of a Th1 response. Indeed, two studies looking at murine models of *Leishmania major* and *Toxoplasma gondii* infection also demonstrate that IL-12 is required for the expansion and the fixation of Th1 cells (Park et al. 2000; Yap et al. 2000).

Distinct functions of the individual members of the IL-12 family not only apply to distinct time points of Th1 cell differentiation or activation. They also apply to *distinct effector mechanisms* initiated by IL-12, IL-23, and IL-27. There is ample evidence for an IL-12/IFN-γ axis (Gately et al. 1998; Trinchieri 1998) (Fig. 4). However, it is clear that murine IL-23 does not regulate IFN-γ production (Cua et al. 2003). Also, human IL-23 induces only modest amounts of IFN-γ in naïve and memory T cells compared to IL-12 (Oppmann et al. 2000). Moreover, IL-23 but not IL-12 can activate murine memory T cells for production of the pro-inflammatory cytokine IL-17 (Aggarwal et al. 2003) (Fig. 4). The activation of IL-17 production by IL-23 appears to be the key mechanism for the emerging role of IL-23 in chronic inflammation (see Sect. 7.7). IL-17 plays an important role in tissue destruction during autoimmune diseases such as multiple sclerosis, rheumatoid arthritis, and psoriasis

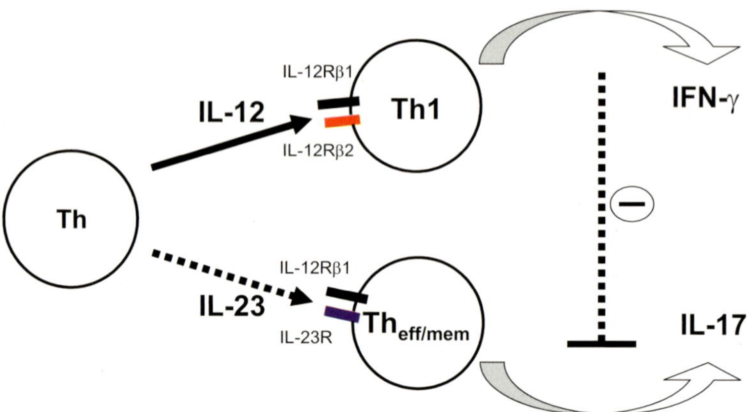

Fig. 4. Th cell differentiation pathways initiated by IL-12 or IL-23. Shown are the IL-12/IFN-γ axis and the IL-23/IL-17 pathway. There is evidence for cross-regulation of these two pro-inflammatory responses (Murphy et al. 2003). The *dashed arrow* indicates that it is presently unclear whether IL-23 drives naïve precursor Th cells or already differentiated Th cells

(Kolls and Linden 2004; Matusevicius et al. 1999; Ziolkowska et al. 2000; Teunissen et al. 1998).

Further evidence for distinct actions of the three IL-12 family members was recently provided by studies shedding light on an unexpected anti-inflammatory potential of IL-27 (Villarino et al. 2003; Hamano et al. 2003). It was found that activation of WSX-1 exerts a powerful negative feedback mechanism that limits T cell hyperactivity and IFN-γ production. The molecular basis for this protective immunoregulatory role of IL-27 might be its ability to activate preferentially STAT1 and STAT3 (Villarino et al. 2004) (Fig. 3).

7.5   Role of IL-12 Family Members in Host Defense

While the role of IL-12 in host defense has been characterized extensively (Gately et al. 1998; Trinchieri 1998; Brombacher et al. 2003), the analysis of the function of IL-23 and IL-27 in resistance and immunity to pathogens is still in the initial phase.

The property of IL-12 to stimulate IFN-γ production during innate and adaptive immune responses is the basis for its essential role in resistance to different types of pathogens. IL-12 has a key role in protection against intracellular protozoan, fungal, and bacterial infections (Brombacher et al. 2003). The role of IL-12 in protection during viral infections appears to be minor (Schijns et al. 1998; Oxenius et al. 1999). For viral infections, type I IFN (IFN-α/β) is of great importance (Guidotti and Chisari 2001). Thus, IL-12 has a pathogen-dependent role in host defense. The data from murine experimental infections were confirmed by the analysis of patients with mutations in the IL-12p40 or the IL-12Rβ1 gene (Ottenhoff et al. 1998). These patients responded normally to standard viral immunizations but developed chronic courses of salmonellosis or mycobacteriosis (Ottenhoff et al. 1998). It is intriguing that IFN-γ deficiency of humans predisposes them primarily against mycobacterial infection (Maclennan et al. 2004). In contrast, patients with IL-12/IL-12Rβ1 deficiency developed salmonellosis (Maclennan et al. 2004). This strongly suggests that IL-12/IL-23 mediate their actions in salmonellosis partly through IFN-γ-independent pathways.

In some murine models of intracellular infection, IL-12p35$^{-/-}$ mice (deficient in IL-12) showed higher resistance than IL-12p35/40$^{-/-}$ mice (deficient in IL-12 *and* IL-23) (Brombacher et al. 2003). This pointed to a role of IL-23 in host responses against these intracellular pathogens. To address this question more directly, very recently infection experiments using IL-23p19$^{-/-}$ have been started. Initial results looking at *T. gondii* infection show a limited role of IL-23 in protection (Lieberman et al. 2004). A significant role of IL-23 (or other p40-dependent proteins such as monomeric or homodimeric p40) in intracellular infection is more evident in the absence of IL-12 (Decken et al. 1998; Lehmann et al. 2001; Lieberman et al. 2004). Presently it is too early to draw final conclusions for the role of IL-23 in innate and adaptive immunity during infection.

First studies looking at the role of IL-27 in host defense were conducted with WSX1/TCCR$^{-/-}$ mice. In *Listeria monocytogenes*, *Leishmania major*, or mycobacterial BCG infection, a role of WSX-1/TCCR signaling in IFN-γ production as well as in granuloma formation was proposed (Chen et al. 2000; Yoshida et al. 2001). However, more recent studies in *Trypanosoma cruzi* or *Toxoplasma gondii* infection show that WSX-1/TCCR is not required for Th1/IFN-γ-mediated immunity to these parasitic infections, but rather for the suppression of excessive production of IFN-γ and T cell hyperactivity (Villarino et al. 2003; Hamano et al. 2003). On the other hand, EBI3$^{-/-}$ mice have a defect in a Th2-mediated colitis model but show an unaltered phenotype in a Th1-mediated colitis model (Nieuwenhuis et al. 2002). This extends the action of IL-27 to Th2 induction (unless EBI-3 could associate with a different subunit than p28, forming a cytokine different from IL-27). Presently it is not clear how these pleotropic effects by IL-27/WSX-1 signaling can be reconciled. It was suggested that the outcome of IL-27/WSX-1 signaling depends on the intensity of Th1 polarization. In a weakly Th1-biased situation such as leishmaniasis, enhancement of early IFN-γ production (Yoshida et al. 2001; Artis et al. 2004) occurs by the IL-27/WSX-1 interaction. However, in extremely polarized Th1 situations such as trypanosomiasis or toxoplasmosis WSX-1 appears to mediate a protective immunosuppressive effect (Villarino et al. 2004).

7.6 Antitumor Activity of IL-12 Family Members

Both endogenous IL-12 and treatment with exogenous IL-12 have been shown to exert profound antitumor and antimetastatic activity in murine models of transplantable and chemically induced tumors (Colombo and Trinchieri 2002). These antitumor activities include innate and adaptive immune mechanisms. In all tumor studies, the antitumor effect of IL-12 was at least partially dependent on IFN-γ (Brunda et al. 1995). IFN-γ and a number of cytokines and chemokines induced by IFN-γ (e.g., IFN-inducible protein 10) have a direct toxic effect on the tumor cells and/or can induce anti-angiogenic mechanisms. Both Th1 cells and cytotoxic T cells contribute to the tumor antigen-specific responses initiated by IL-12. This results in protective immunity to challenge with the same type of tumor (Brunda et al. 1993) and might provide the basis for potential tumor-specific vaccination using IL-12 as an adjuvant. The induction of Th1 cells by IL-12 not only results in elevated IFN-γ production, but also in higher levels of opsonizing and complement-fixing IgG antibody isotypes that have been demonstrated to contribute to antitumor responses (Quaglino et al. 2002). Unfortunately, the induction of efficient in vivo antitumor responses requires high doses of recombinant IL-12. Since IL-12 is a pro-inflammatory cytokine, high-dose treatment led to considerable toxicity in mice and humans (Trinchieri 1998). It remains a challenge to translate the promising preclinical data obtained with IL-12 immunotherapy to human cancer patients and at the same time minimize the toxicity by IL-12 treatment.

The antitumor potential of IL-23 was recently started to be studied in models of colon carcinoma cells and melanoma cells (Lo et al. 2003; Wang et al. 2003). Potent activities were found in these initial studies. CD8$^+$ T cells but not CD4$^+$ T cells or NK cells were found to be required for the IL-23-induced antitumor effects and the induction of systemic immunity. Interestingly, compared with IL-12-expressing tumor cells, IL-23-transduced tumor cells lacked an early host response but achieved comparable antitumor and antimetastatic activity (Lo et al. 2003). This indicates that IL-23 displays some antitumor mechanisms that are distinct from the actions of IL-12.

Very recently, a first study described the antitumor activity of IL-27 in a murine tumor model of colon carcinoma. Potent protective immunity induced by IL-27 was found (Hisada et al. 2004). The antitumor response depended on CD8$^+$ T cells, IFN-γ, T-bet but not on STAT4.

Taken together, all members of the IL-12 family appear to have profound effects in tumor control. For the future, clinical trials will have to reveal how the members of the IL-12 family can be used for treatment of cancer patients without inducing major toxicities.

7.7 Role of IL-12 Family Members in Organ-Specific Autoimmunity

Historically, IFN-γ producing Th1 cells induced by IL-12 were considered to be mediators of autoimmunity. These studies aiming to look at the role of IL-12 in autoimmunity were in many cases done using neutralizing anti-p40 antibodies, p40$^{-/-}$ or IL-12Rβ1$^{-/-}$ mice. With today's knowledge on p40 being shared between IL-12 and IL-23 and that on IL-12Rβ1 being used by both IL-12 and IL-23, it has become clear that any conclusions from these studies leave open whether IL-12 and/or IL-23 were responsible for the observed effects. More specific information on the role of IL-12 can be derived from the analysis of mice deficient in IL-12p35 or IL-12Rβ2. On the other hand, the recent generation of IL-23p19$^{-/-}$ mice has allowed for definitive analysis of the role of IL-23 in autoimmunity (Cua et al. 2003; Murphy et al. 2003).

7.7.1   Experimental Autoimmune Encephalomyelitis

Experimental autoimmune encephalomyelitis (EAE), a murine model of multiple sclerosis, is considered to be a Th1-mediated autoimmune disease. EAE could not be induced in IL-12p40$^{-/-}$ mice, but was even more severe in IL-12p35$^{-/-}$ and IL-12Rβ2$^{-/-}$ mice than in wild-type mice (Gran et al. 2002; Zhang et al. 2003). These data exclude an essential role of IL-12 in EAE and instead point to IL-23. Indeed, IL-23p19$^{-/-}$ mice were found to be completely resistant to EAE induction (Cua et al. 2003). IL-23p19$^{-/-}$ mice showed no symptoms of EAE but normal induction of Th1 cells. It is noteworthy that the generation of a Th1 response to myelin oligodendrocyte glycoprotein in this model required IL-12 and not IL-23. Intracerebral expression of IL-23 in IL-23p19$^{-/-}$ or IL-12p40$^{-/-}$ mice reconstituted EAE. However, IL-23 treatment of IL-12p40$^{-/-}$ mice led to delayed disease and reduced disease severity as compared to IL-23 treatment of IL-23p19$^{-/-}$ mice (Cua et al. 2003). In IL-12p40$^{-/-}$ mice, prior intraperitoneal administration of IL-12 and subsequent intracerebral expression of IL-23 by gene transfer resulted in disease severity similar to wild-type mice. This suggests that IL-23 is essential for EAE by local inflammatory effector mechanisms and IL-12 contributes to disease expression by promoting Th1 development. Since the IL-23R components were expressed only on recruited inflammatory macrophages, direct activation of macrophages by IL-23 appears to be responsible for the CNS inflammation during EAE. In addition, more recent data shows that IL-23p19$^{-/-}$ mice are severely impaired in their capacity to develop IL-17 producing Th effector/memory cells (Murphy et al. 2003; Ghilardi et al. 2004). IL-17 production by memory Th cells has been found earlier to be induced by IL-23 (Aggarwal et al. 2003) (Fig. 4). Upon transfer of IL-17 producing wild-type Th cells from mice with EAE to naïve recipient mice, the pathogenic IL-23-driven IL-17-producing Th cells can invade the CNS and promote EAE (Langrish et al. 2005).

7.7.2 Rheumatoid Arthritis

In collagen-induced arthritis (CIA), a murine model of rheumatoid arthritis (RA), IL-23p19$^{-/-}$ mice displayed resistance, whereas IL-12p35$^{-/-}$ mice were highly susceptible (i.e., more susceptible than wild-type mice), suggesting an unexpected suppressive action of IL-12 on chronic joint inflammation (Murphy et al. 2003). Again, in addition to the function of IL-23 vs IL-12 in chronic inflammation of the CNS, IL-23 but not IL-12 is essential for development of joint autoimmune inflammation. Moreover, IL-23p19$^{-/-}$ mice were resistant in CIA despite developing collagen-specific, IFN-γ-producing Th1 cells (Murphy et al. 2003). Murphy et al. reported that lymph node cells from antigen-primed wild-type and IL-12p35$^{-/-}$ mice have an increased frequency of IL-17-producing CD4$^+$ T cells as compared with IL-23p19$^{-/-}$ mice (Murphy et al. 2003). Therefore, IL-17 production depends on IL-23 in CIA as described above for EAE. The more severe course of CIA in IL-12p35$^{-/-}$ mice than in wild-type mice was associated with elevated IL-17 production in IL-12p35$^{-/-}$ mice vs wild-type mice. These data for the first time suggest that the IL-23/IL-17 pathway is negatively regulated by the IL-12/IFN-γ axis (Fig. 4). Recently in IL-17$^{-/-}$ mice suppression of CIA was described confirming the central role of IL-17 in RA (Nakae et al. 2003). A role for IL-17 in human autoimmune tissue destruction has been suggested before by others who found IL-17 expression in several human autoimmune diseases, including multiple sclerosis (Matusevicius et al. 1999), RA (Ziolkowska et al. 2000), and psoriasis (Teunissen et al. 1998).

7.7.3 Psoriasis Vulgaris

Psoriasis vulgaris is a T cell-driven disease with Th1 cells predominating in lesional skin (Austin et al. 1999). Keratinocytes with transgenic expression of p40 show production of IL-23, but not IL-12, in the skin leading to an inflammatory cutaneous response with elevated numbers of Langerhans' cells (Kopp et al. 2003). Moreover, in lesional skin of patients with psoriasis vulgaris, an increase of p40 and p19 but not of p35 mRNA compared with nonlesional skin was detected (Lee et al. 2004). This emphasizes a major role of IL-23 in cutaneous chronic inflammatory responses.

7.7.4 Inflammatory Bowel Disease

Inflammatory bowel disease (IBD) encompasses two disease entitities, Crohn's disease and ulcerative colitis. Crohn's disease most commonly affects the terminal ileum and ascending colon. An immunopathological function of IL-12 in IBD was assumed for Crohn's disease, since patients with Crohn's disease were found to express IL-12 in the gut (Parronchi et al. 1997). In addition, in mouse models of IBD, treatment with anti-IL-12p40 antibodies prevented or terminated disease (Neurath et al. 1995; Davidson et al. 1998). Interestingly, treatment with anti-IFN-γ could suppress induction of IBD but was unable to reverse ongoing IBD (Davidson et al. 1998). Based on these outcomes, it was concluded that IL-12 might play a role in colitis independent of its ability to generate IFN-γ-producing T cells. With the discovery of IL-23, the role of IL-12 in IBD needed to be readdressed, since anti-p40 antibodies used before neutralized both IL-12 and IL-23. Indeed, very recent studies in IL-23p19$^{-/-}$ mice revealed an essential role of IL-23 in development of IBD (Yen et al., unpublished). Also, IL-23 expression was found in patients with active Crohn's disease (Stallmach et al. 2004). Moreover, murine lamina propria dendritic cells were found to express constitutively IL-23 in the terminal ileum driven by the intestinal flora (Becker et al. 2003). Interestingly, IL-23 expression in this part of the small intestine was accompanied by expression of IL-17 (Becker et al. 2003). This suggests that IL-23 might predispose the terminal ileum to initiate chronic inflammatory responses mediated by IL-17. These data place the IL-23/IL-17 pathway in the center of the manifestation of chronic intestinal inflammation. The IL-12/IFN-γ axis might contribute to the induction and progression of IBD but is not essential.

Presently only few data are available on the potential role of IL-27 in IBD. Expression of EBI3, one of the subunits of IL-27, was found in patients with ulcerative colitis and in a subgroup of patients with Crohn's disease (Omata et al. 2001). Since a major anti-inflammatory function of IL-27 has been described (Villarino et al. 2004), it will be exciting to define a potential role of IL-27 in IBD and other autoimmune diseases.

7.7.5 Insulin-Dependent Diabetes Mellitus

For insulin-dependent diabetes mellitus (IDDM), some evidence was provided for an involvement of IL-12. A role of IL-12 in IDDM is based on the pancreatic expression of IL-12p40 found in several studies (Rothe et al. 1996; Zipris et al. 1996). In NOD (nonobese diabetes) mice, which are genetically predisposed to the development of IDDM, IL-12 treatment accelerated the onset of disease (Trembleau et al. 1995). However, using a different protocol of IL-12 administration, suppression of IDDM development in NOD mice was found (O'Hara et al. 1996). Also, IL-12p40$^{-/-}$ mice backcrossed to the NOD background showed a reduced incidence of IDDM (Adorini et al. 1997). However, these mice are also unable to produce IL-23. Therefore, studies with IL-12p35$^{-/-}$ (lacking only IL-12) and IL-23p19$^{-/-}$ (lacking only IL-23) should allow for clarification of the individual roles of IL-12 vs IL-23 in the pathogenesis of IDDM.

Acknowledgements. The authors are grateful for financial support of their research by the Deutsche Forschungsgemeinschaft (AL 371/3–3), Friedrich-Ebert-Stiftung, Hanns-Seidel-Stiftung, and Schaumann Stiftung.

References

Adorini L, Aloisi F, Galbiati F, Gately MK, Gregori S, Penna G, Ria F, Smiroldo S, Trembleau S (1997) Targeting IL-12, the key cytokine driving Th1-mediated autoimmune diseases. Chem Immunol 68:175–197

Aggarwal S, Ghilardi N, Xie MH, De Sauvage FJ, Gurney AL (2003) Interleukin-23 promotes a distinct CD4 T cell activation state characterized by the production of interleukin-17. J Biol Chem 278:1910–1914

Artis D, Johnson LM, Joyce K, Saris C, Villarino A, Hunter CA, Scott P (2004) Cutting edge: early IL-4 production governs the requirement for IL-27-WSX-1 signaling in the development of protective Th1 cytokine responses following Leishmania major infection. J Immunol 172:4672–4675

Austin LM, Ozawa M, Kikuchi T, Walters IB, Krueger JG (1999) The majority of epidermal T cells in Psoriasis vulgaris lesions can produce type 1 cytokines, interferon-gamma, interleukin-2, and tumor necrosis factor-alpha, defining TC1 (cytotoxic T lymphocyte) and TH1 effector populations: a type 1 differentiation bias is also measured in circulating blood T cells in psoriatic patients. J Invest Dermatol 113:752–759

Becker C, Wirtz S, Blessing M, Pirhonen J, Strand D, Bechthold O, Frick J, Galle PR, Autenrieth I, Neurath MF (2003) Constitutive p40 promoter activation and IL-23 production in the terminal ileum mediated by dendritic cells. J Clin Invest 112:693–706

Belladonna ML, Renauld JC, Bianchi R, Vacca C, Fallarino F, Orabona C, Fioretti MC, Grohmann U, Puccetti P (2002) IL-23 and IL-12 have overlapping, but distinct, effects on murine dendritic cells. J Immunol 168:5448–5454

Brombacher F, Kastelein RA, Alber G (2003) Novel IL-12 family members shed light on the orchestration of Th1 responses. Trends Immunol 24:207–212

Brunda MJ, Luistro L, Hendrzak JA, Fountoulakis M, Garotta G, Gately MK (1995) Role of interferon-gamma in mediating the antitumor efficacy of interleukin-12. J Immunother Emphasis Tumor Immunol 17:71–77

Brunda MJ, Luistro L, Warrier RR, Wright RB, Hubbard BR, Murphy M, Wolf SF, Gately MK (1993) Antitumor and antimetastatic activity of interleukin 12 against murine tumors. J Exp Med 178:1223–1230

Chen Q, Ghilardi N, Wang H, Baker T, Xie MH, Gurney A, Grewal IS, De Sauvage FJ (2000) Development of Th1-type immune responses requires the type I cytokine receptor TCCR. Nature 407:916–920

Colombo MP, Trinchieri G (2002) Interleukin-12 in anti-tumor immunity and immunotherapy. Cytokine Growth Factor Rev 13:155–168

Cua DJ, Sherlock J, Chen Y, Murphy CA, Joyce B, Seymour B, Lucian L, To W, Kwan S, Churakova T, Zurawski S, Wiekowski M, Lira SA, Gorman D, Kastelein RA, Sedgwick JD (2003) Interleukin-23 rather than interleukin-12 is the critical cytokine for autoimmune inflammation of the brain. Nature 421:744–748

D'Andrea A, Rengaraju M, Valiante NM, Chehimi J, Kubin M, Aste M, Chan SH, Kobayashi M, Young D, Nickbarg E et al (1992) Production of natural killer cell stimulatory factor (interleukin 12) by peripheral blood mononuclear cells. J Exp Med 176:1387–1398

Davidson NJ, Hudak SA, Lesley RE, Menon S, Leach MW, Rennick DM (1998) IL-12, but not IFN-gamma, plays a major role in sustaining the chronic phase of colitis in IL-10-deficient mice. J Immunol 161:3143–3149

Decken K, Kohler G, Palmer-Lehmann K, Wunderlin A, Mattner F, Magram J, Gately MK, Alber G (1998) Interleukin-12 is essential for a protective Th1 response in mice infected with Cryptococcus neoformans. Infect Immun 66:4994–5000

Gately MK, Renzetti LM, Magram J, Stern AS, Adorini L, Gubler U, Presky DH (1998) The interleukin-12/interleukin-12-receptor system: role in normal and pathologic immune responses. Annu Rev Immunol 16:495–521

Gearing DP, Cosman D (1991) Homology of the p40 subunit of natural killer cell stimulatory factor (NKSF) with the extracellular domain of the interleukin-6 receptor. Cell 66:9–10

Ghilardi N, Kljavin N, Chen Q, Lucas S, Gurney AL, De Sauvage FJ (2004) Compromised humoral and delayed-type hypersensitivity responses in IL-23-deficient mice. J Immunol 172:2827–2833

Gran B, Zhang GX, Yu S, Li J, Chen XH, Ventura ES, Kamoun M, Rostami A (2002) IL-12p35-deficient mice are susceptible to experimental autoimmune encephalomyelitis: evidence for redundancy in the IL-12 system in the induction of central nervous system autoimmune demyelination. J Immunol 169:7104–7110

Grohmann U, Belladonna ML, Bianchi R, Orabona C, Ayroldi E, Fioretti MC, Puccetti P (1998) IL-12 acts directly on DC to promote nuclear localization of NF-kappaB and primes DC for IL-12 production. Immunity 9:315–323

Grohmann U, Belladonna ML, Vacca C, Bianchi R, Fallarino F, Orabona C, Fioretti MC, Puccetti P (2001) Positive regulatory role of IL-12 in macrophages and modulation by IFN-gamma. J Immunol 167:221–227

Gubler U, Chua AO, Schoenhaut DS, Dwyer CM, McComas W, Motyka R, Nabavi N, Wolitzky AG, Quinn PM, Familletti PC (1991) Coexpression of two distinct genes is required to generate secreted bioactive cytotoxic lymphocyte maturation factor. Proc Natl Acad Sci U S A 88:4143–4147

Guidotti LG, Chisari FV (2001) Noncytolytic control of viral infections by the innate and adaptive immune response. Annu Rev Immunol 19:65–91

Hamano S, Himeno K, Miyazaki Y, Ishii K, Yamanaka A, Takeda A, Zhang M, Hisaeda H, Mak TW, Yoshimura A, Yoshida H (2003) WSX-1 is required for resistance to Trypanosoma cruzi infection by regulation of proinflammatory cytokine production. Immunity 19:657–667

Hisada M, Kamiya S, Fujita K, Belladonna ML, Aoki T, Koyanagi Y, Mizuguchi J, Yoshimoto T (2004) Potent antitumor activity of interleukin-27. Cancer Res 64:1152–1156

Kaplan MH, Sun YL, Hoey T, Grusby MJ (1996) Impaired IL-12 responses and enhanced development of Th2 cells in Stat4-deficient mice. Nature 382:174–177

Kobayashi M, Fitz L, Ryan M, Hewick RM, Clark SC, Chan S, Loudon R, Sherman F, Perussia B, Trinchieri G (1989) Identification and purification of natural killer cell stimulatory factor (NKSF), a cytokine with multiple biologic effects on human lymphocytes. J Exp Med 170:827–845

Kolls JK, Linden A (2004) Interleukin-17 family members and inflammation. Immunity 21:467–476

Kopp T, Lenz P, Bello-Fernandez C, Kastelein RA, Kupper TS, Stingl G (2003) IL-23 production by cosecretion of endogenous p19 and transgenic p40 in keratin 14/p40 transgenic mice: evidence for enhanced cutaneous immunity. J Immunol 170:5438–5444

Langrish CL, Chen Y, Blumenschein WM, Mattson J, Basham B, Sedgwick JD, McClanahan T, Kastelein RA, Cua DJ (2005) IL-23 drives a pathogenic T cell population that induces autoimmune inflammation. J Exp Med 201:233–240

Lee E, Trepicchio WL, Oestreicher JL, Pittman D, Wang F, Chamian F, Dhodapkar M, Krueger JG (2004) Increased expression of interleukin 23 p19 and p40 in lesional skin of patients with psoriasis vulgaris. J Exp Med 199:125–130

Lehmann J, Bellmann S, Werner C, Schroder R, Schutze N, Alber G (2001) IL-12p40-dependent agonistic effects on the development of protective innate and adaptive immunity against Salmonella enteritidis. J Immunol 167:5304–5315

Lieberman LA, Cardillo F, Owyang AM, Rennick DM, Cua DJ, Kastelein RA, Hunter CA (2004) IL-23 provides a limited mechanism of resistance to acute toxoplasmosis in the absence of IL-12. J Immunol 173:1887–1893

Lo CH, Lee SC, Wu PY, Pan WY, Su J, Cheng CW, Roffler SR, Chiang BL, Lee CN, Wu CW, Tao MH (2003) Antitumor and antimetastatic activity of IL-23. J Immunol 171:600–607

Maclennan C, Fieschi C, Lammas DA, Picard C, Dorman SE, Sanal O, Maclennan JM, Holland SM, Ottenhoff TH, Casanova JL, Kumararatne DS (2004) Interleukin (IL)-12 and IL-23 are key cytokines for immunity against salmonella in humans. J Infect Dis 190:1755–1757

Matusevicius D, Kivisakk P, He B, Kostulas N, Ozenci V, Fredrikson S, Link H (1999) Interleukin-17 mRNA expression in blood and CSF mononuclear cells is augmented in multiple sclerosis. Mult Scler 5:101–104

Merberg DM, Wolf SF, Clark SC (1992) Sequence similarity between NKSF and the IL-6/G-CSF family. Immunol Today 13:77–78

Mullen AC, High FA, Hutchins AS, Lee HW, Villarino AV, Livingston DM, Kung AL, Cereb N, Yao TP, Yang SY, Reiner SL (2001) Role of T-bet in commitment of TH1 cells before IL-12-dependent selection. Science 292:1907–1910

Murphy CA, Langrish CL, Chen Y, Blumenschein W, McClanahan T, Kastelein RA, Sedgwick JD, Cua DJ (2003) Divergent pro- and antiinflammatory roles for IL-23 and IL-12 in joint autoimmune inflammation. J Exp Med 198:1951–1957

Nakae S, Nambu A, Sudo K, Iwakura Y (2003) Suppression of immune induction of collagen-induced arthritis in IL-17-deficient mice. J Immunol 171:6173–6177

Neurath MF, Fuss I, Kelsall BL, Stuber E, Strober W (1995) Antibodies to interleukin 12 abrogate established experimental colitis in mice. J Exp Med 182:1281–1290

Nieuwenhuis EE, Neurath MF, Corazza N, Iijima H, Trgovcich J, Wirtz S, Glickman J, Bailey D, Yoshida M, Galle PR, Kronenberg M, Birkenbach M, Blumberg RS (2002) Disruption of T helper 2-immune responses in Epstein-Barr virus-induced gene 3-deficient mice. Proc Natl Acad Sci U S A 99:16951–16956

O'Hara RM Jr, Henderson SL, Nagelin A (1996) Prevention of a Th1 disease by a Th1 cytokine: IL-12 and diabetes in NOD mice. Ann N Y Acad Sci 795:241–249

Omata F, Birkenbach M, Matsuzaki S, Christ AD, Blumberg RS (2001) The expression of IL-12 p40 and its homologue, Epstein-Barr virus-induced gene 3, in inflammatory bowel disease. Inflamm Bowel Dis 7:215–220

Oppmann B, Lesley R, Blom B, Timans JC, Xu Y, Hunte B, Vega F, Yu N, Wang J, Singh K, Zonin F, Vaisberg E, Churakova T, Liu M, Gorman D, Wagner J, Zurawski S, Liu Y, Abrams JS, Moore KW, Rennick D, Waal-Malefyt R, Hannum C, Bazan JF, Kastelein RA (2000) Novel p19 protein engages IL-12p40 to form a cytokine, IL-23, with biological activities similar as well as distinct from IL-12. Immunity 13:715–725

Ottenhoff TH, Kumararatne D, Casanova JL (1998) Novel human immunodeficiencies reveal the essential role of type-I cytokines in immunity to intracellular bacteria. Immunol Today 19:491–494

Oxenius A, Karrer U, Zinkernagel RM, Hengartner H (1999) IL-12 is not required for induction of type 1 cytokine responses in viral infections. J Immunol 162:965–973

Parham C, Chirica M, Timans J, Vaisberg E, Travis M, Cheung J, Pflanz S, Zhang R, Singh KP, Vega F, To W, Wagner J, O'Farrell AM, McClanahan T, Zurawski S, Hannum C, Gorman D, Rennick DM, Kastelein RA, de Waal MR, Moore KW (2002) A receptor for the heterodimeric cytokine IL-23 is composed of IL-12Rbeta1 and a novel cytokine receptor subunit, IL-23R. J Immunol 168:5699–5708

Park AY, Hondowicz BD, Scott P (2000) IL-12 is required to maintain a Th1 response during Leishmania major infection. J Immunol 165:896–902

Parronchi P, Romagnani P, Annunziato F, Sampognaro S, Becchio A, Giannarini L, Maggi E, Pupilli C, Tonelli F, Romagnani S (1997) Type 1 T-helper cell predominance and interleukin-12 expression in the gut of patients with Crohn's disease. Am J Pathol 150:823–832

Pflanz S, Hibbert L, Mattson J, Rosales R, Vaisberg E, Bazan JF, Phillips JH, McClanahan TK, de Waal MR, Kastelein RA (2004) WSX-1 and glycoprotein 130 constitute a signal-transducing receptor for IL-27. J Immunol 172:2225–2231

Pflanz S, Timans JC, Cheung J, Rosales R, Kanzler H, Gilbert J, Hibbert L, Churakova T, Travis M, Vaisberg E, Blumenschein WM, Mattson JD, Wagner JL, To W, Zurawski S, McClanahan TK, Gorman DM, Bazan JF, de Waal MR, Rennick D, Kastelein RA (2002) IL-27, a heterodimeric cytokine composed of EBI3 and p28 protein, induces proliferation of naive CD4(+) T cells. Immunity 16:779–790

Presky DH, Yang H, Minetti LJ, Chua AO, Nabavi N, Wu CY, Gately MK, Gubler U (1996) A functional interleukin 12 receptor complex is composed of two beta-type cytokine receptor subunits. Proc Natl Acad Sci U S A 93:14002–14007

Quaglino E, Rovero S, Cavallo F, Musiani P, Amici A, Nicoletti G, Nanni P, Forni G (2002) Immunological prevention of spontaneous tumors: a new prospect? Immunol Lett 80:75–79

Rothe H, Burkart V, Faust A, Kolb H (1996) Interleukin-12 gene expression is associated with rapid development of diabetes mellitus in non-obese diabetic mice. Diabetologia 39:119–122

Schijns VE, Haagmans BL, Wierda CM, Kruithof B, Heijnen IA, Alber G, Horzinek MC (1998) Mice lacking IL-12 develop polarized Th1 cells during viral infection. J Immunol 160:3958–3964

Schulz O, Edwards AD, Schito M, Aliberti J, Manickasingham S, Sher A, Reis e Sousa C (2000) CD40 triggering of heterodimeric IL-12 p70 production by dendritic cells in vivo requires a microbial priming signal. Immunity 13:453–462

Stallmach A, Giese T, Schmidt C, Ludwig B, Mueller-Molaian I, Meuer SC (2004) Cytokine/chemokine transcript profiles reflect mucosal inflammation in Crohn's disease. Int J Colorectal Dis 19:308–315

Stern AS, Podlaski FJ, Hulmes JD, Pan YC, Quinn PM, Wolitzky AG, Familletti PC, Stremlo DL, Truitt T, Chizzonite R et al (1990) Purification to homogeneity and partial characterization of cytotoxic lymphocyte maturation factor from human B-lymphoblastoid cells. Proc Natl Acad Sci U S A 87:6808–6812

Takeda A, Hamano S, Yamanaka A, Hanada T, Ishibashi T, Mak TW, Yoshimura A, Yoshida H (2003) Cutting edge: role of IL-27/WSX-1 signaling for induction of T-bet through activation of STAT1 during initial Th1 commitment. J Immunol 170:4886–4890

Teunissen MB, Koomen CW, de Waal MR, Wierenga EA, Bos JD (1998) Interleukin-17 and interferon-gamma synergize in the enhancement of proinflammatory cytokine production by human keratinocytes. J Invest Dermatol 111:645–649

Trembleau S, Penna G, Bosi E, Mortara A, Gately MK, Adorini L (1995) Interleukin 12 administration induces T helper type 1 cells and accelerates autoimmune diabetes in NOD mice. J Exp Med 181:817–821

Trinchieri G (1998) Interleukin-12: a cytokine at the interface of inflammation and immunity. Adv Immunol 70:83–243

Trinchieri G (2003) Interleukin-12 and the regulation of innate resistance and adaptive immunity. Nat Rev Immunol 3:133–146

Villarino A, Hibbert L, Lieberman L, Wilson E, Mak T, Yoshida H, Kastelein RA, Saris C, Hunter CA (2003) The IL-27R (WSX-1) is required to suppress T cell hyperactivity during infection. Immunity 19:645–655

Villarino AV, Huang E, Hunter CA (2004) Understanding the pro- and anti-inflammatory properties of IL-27. J Immunol 173:715–720

Wang YQ, Ugai S, Shimozato O, Yu L, Kawamura K, Yamamoto H, Yamaguchi T, Saisho H, Tagawa M (2003) Induction of systemic immunity by expression of interleukin-23 in murine colon carcinoma cells. Int J Cancer 105:820–824

Yap G, Pesin M, Sher A (2000) Cutting edge: IL-12 is required for the maintenance of IFN-gamma production in T cells mediating chronic resistance to the intracellular pathogen, Toxoplasma gondii. J Immunol 165:628–631

Yoshida H, Hamano S, Senaldi G, Covey T, Faggioni R, Mu S, Xia M, Wakeham AC, Nishina H, Potter J, Saris CJ, Mak TW (2001) WSX-1 is required for the initiation of Th1 responses and resistance to L. major infection. Immunity 15:569–578

Zhang GX, Gran B, Yu S, Li J, Siglienti I, Chen X, Kamoun M, Rostami A (2003) Induction of experimental autoimmune encephalomyelitis in IL-12 receptor-beta 2-deficient mice: IL-12 responsiveness is not required in the pathogenesis of inflammatory demyelination in the central nervous system. J Immunol 170:2153–2160

Ziolkowska M, Koc A, Luszczykiewicz G, Ksiezopolska-Pietrzak K, Klimczak E, Chwalinska-Sadowska H, Maslinski W (2000) High levels of IL-17 in rheumatoid arthritis patients: IL-15 triggers in vitro IL-17 production via cyclosporin A-sensitive mechanism. J Immunol 164:2832–2838

Zipris D, Greiner DL, Malkani S, Whalen B, Mordes JP, Rossini AA (1996) Cytokine gene expression in islets and thyroids of BB rats. IFN-gamma and IL-12p40 mRNA increase with age in both diabetic and insulin-treated nondiabetic BB rats. J Immunol 156:1315–1321

8 The Role of TNFα and IL-17 in the Development of Excess IL-1 Signaling-Induced Inflammatory Diseases in IL-1 Receptor Antagonist-Deficient Mice

H. Ishigame, A. Nakajima, S. Saijo, Y. Komiyama, A. Nambu,
T. Matsuki, S. Nakae, R. Horai, S. Kakuta, Y. Iwakura

8.1   Introduction 130
8.2   The Roles of TNFα and IL-17 in the Development of Arthritis . . . 133
8.2.1   Development of Autoimmune Arthritis in IL-1Ra$^{-/-}$ Mice 133
8.2.2   The Role of TNFα in the Development of Arthritis 135
8.2.3   The Role of IL-17 in the Development of Arthritis 138
8.3   The Roles of TNFα and IL-17 in the Development of Aortitis . . . 140
8.3.1   Development of Aortitis in IL-1Ra$^{-/-}$ Mice 140
8.3.2   The Roles of TNFα and IL-17 in the Development of Aortitis . . . 142
8.4   The Role of TNFα in the Development of Dermatitis 144
8.4.1   Development of Psoriasis-Like Dermatitis in IL-1Ra$^{-/-}$ Mice . . . 144
8.4.2   The Role of TNFα in the Development of Dermatitis 145
8.5   Conclusion 146
References 147

Abstract. IL-1 receptor antagonist (IL-1Ra)-deficient mice spontaneously develop several inflammatory diseases, resembling rheumatoid arthritis, aortitis, and psoriasis in humans. As adoptive T cell transplantation could induce arthritis and aortitis in recipient mice, it was suggested that an autoimmune process is involved in the development of diseases. In contrast, as dermatitis developed in *scid/scid*-IL-1Ra-deficient mice and could not be induced by T cell transfer, a T cell-independent mechanism was suggested. The expression of proinflammatory cytokines was augmented at the inflammatory sites. The development of arthritis and aortitis was significantly suppressed by the deficiency of TNFα or IL-17. The development of dermatitis was also inhibited by the deficiency of TNFα. These observations suggest that TNFα and IL-17 play a crucial role in the development of autoimmunity downstream of IL-1 signaling, and excess IL-1 signaling-induced TNFα also induces skin inflammation in a T cell-independent manner.

8.1 Introduction

IL-1 is a proinflammatory cytokine functioning in inflammation and host responses to infection (for reviews, see Durum and Oppenheim 1993; Dinarello 1996; Tocci and Schmidt 1997; Nakae et al. 2003a). Originally identified as an endogenous pyrogen, the alternate names for IL-1 of lymphocyte-activating factor, hemopoietin-1, and osteoclast activating factor serve to demonstrate its pleiotropic activity (Dinarello 1991). IL-1, produced by a variety of cells including macrophages, monocytes, keratinocytes, and synovial lining cells, induces inflammation via the activation of synovial cells, endothelial cells, lymphocytes, and macrophages. Upon activation, these cells produce a variety of additional chemokines, cytokines, and inflammatory mediators (Feldmann et al. 1996), including IL-1 itself, IL-17, TNFα, IL-6, IL-8, and cyclooxygenase (COX)-2; these molecules ultimately cause infiltration of leukocytes into inflammatory sites, increase the permeability of blood vessels, and induce fever (Davis and MacIntyre 1992; Dinarello 1996; Tocci and Schmidt 1997; Nakae et al. 2003d).

Two molecular species of IL-1, IL-1α and IL-1β, are derived from two distinct genes on chromosome 2 (mouse, human). Despite a minimal 25% amino acid sequence identity between the molecules (Dinarello 1991), these species exert similar, but not completely overlapping, biological activities through binding to the IL-1 type I receptor (IL-1RI) (Sims et al. 1993; Nakae et al. 2001a, 2001c). Recently, it was reported that the IL-1α precursor, but not IL-1β, moves to the nucleus and activates the transcription of cytokine genes via activation of NF-κB and

AP-1 in an IL-1R-independent mechanism (Werman et al. 2004). While an IL-1 type II receptor (IL-1RII) also exists, this receptor does not appear to function in signal transduction (Colotta et al. 1993).

The IL-1 receptor antagonist (Ra), an additional member of the IL-1 gene family, binds IL-1 receptors without exerting agonist activity (Carter et al. 1990; Hannum et al. 1990). This is because IL-1Ra cannot recruit the IL-1R-accessory protein, a necessary component of an active receptor complex (Greenfeder et al. 1995). Since IL-1Ra competes with IL-1α and IL-1β for the binding of IL-1 receptors (Carter et al. 1990; Hannum et al. 1990), IL-1Ra is considered to be a negative regulator of IL-1 signal.

We have generated IL-1Ra-deficient (IL-1Ra$^{-/-}$) mice (Horai et al. 1998) in which all the three isoforms of IL-1Ra (Muzio et al. 1995) are deleted and demonstrated that these mice spontaneously developed chronic inflammatory arthropathy (Horai et al. 2000). Mice deficient in the IL-1Ra gene have also been reported by Hirsch et al. (1996) and Nicklin et al. (2000); these animals exhibited early mortality and arteritis, respectively. It was recently reported that these mice also spontaneously develop inflammatory dermatitis (Shepherd et al. 2004). However, the pathogenesis of these diseases has not been elucidated completely. Hence, it is remarkable that expression of a variety of inflammatory cytokines, including TNFα, is augmented in these mice.

TNFα, a proinflammatory cytokine, was originally identified as an endotoxin-induced serum factor that causes tumor necrosis (Carswell et al. 1975). TNFα is produced by multiple cell types, including monocytes, macrophages, keratinocytes, and activated T cells. Upon activation with soluble bacterial components or by direct contact with activated T cells at inflammatory sites, TNFα is synthesized in these cells as a membrane-bound precursor. Cleavage by metalloproteinase, TACE/ADAM-17, results in the secretion of a soluble, mature form (Fowlkes and Winkler 2002). While the soluble form may be more potent, both forms of TNFα are biologically active. TNFα can bind to two different cell surface receptors, TNFRI (p55) and TNFRII (p75), on the target cells, including T cells, NK cells, keratinocytes, osteoclasts, and endothelial cells.

IL-17 is another proinflammatory cytokine, originally named cytotoxic T lymphocyte associated serine esterase (CTLA-8) (Rouvier et al.

1993). IL-17 is produced by $TCR\alpha/\beta^+CD4^-CD8^-$ thymocytes, as well as activated $CD4^+$ and $CD4^+CD45RO^+$ memory T cells (Yao et al. 1995b; Kennedy et al. 1996). In humans, activated $CD8^+$ and $CD8^+$ $CD45RO^+$ memory T cells may also produce IL-17 upon activation with PMA/ionomycin (Shin et al. 1999). While Aarvak et al. reported that IL-17 is produced by Th1/Th0 clone cells, but not by Th2 cells found in the joints of rheumatoid arthritis (RA) patients, Albanesi et al. reported that both Th1 and Th2 cell clones from human skin-derived nickel-specific T cells could produce IL-17 (Aarvak et al. 1999; Albanesi et al. 2000). In patients with lyme arthritis and in mice with microbial infection, however, IL-17 is produced by $CD4^+$ T cells expressing $TNF\alpha$, but not by Th1 or Th2 cells (Infante-Duarte et al. 2000). In addition to $CD4^+$ and $CD8^+$ T cells, neutrophils also produce IL-17 upon lipopolysaccharide-induced airway neutrophilia (Ferretti et al. 2003). Therefore, a wide variety of cells of the immune system are capable of producing IL-17 under different conditions.

IL-17 exerts pleiotropic activities through activation of IL-17R, which exhibits a ubiquitous tissue distribution (Yao et al. 1995a). The activities include the induction of $TNF\alpha$, IL-1β, IL-6, IL-8, G-CSF, and MCP-1 on various cell types, the upregulation of ICAM-1 and HLA-DR on keratinocytes, the induction of iNOS and COX-2 on chondrocytes, the stimulation of osteoclast differentiation factor (ODF) production on osteoblasts, and the promotion of SCF- and G-CSF-mediated granulopoiesis (Aggarwal and Gurney 2002; Moseley et al. 2003). IL-17 is detectable in the sera and the diseased organs and tissues of various patients, suggesting involvement in the development of human diseases such as RA, osteoarthritis, multiple sclerosis, systemic lupus erythematosus, and asthma (Aggarwal and Gurney 2002; Kolls and Linden 2004). Furthermore, the use of IL-17R$^{-/-}$ mice has implicated IL-17 in the host defense mechanisms against *Klebsiella pneumoniae* infection (Ye et al. 2001). Recently, we have shown that IL-17 is involved in contact, delayed-type, and airway hypersensitivity responses as well as T-dependent antibody production, but not in acute graft-versus-host reaction, using IL-17$^{-/-}$ mice (Nakae et al. 2002). Furthermore, it was suggested that impaired responses were caused by the defects of allergen-specific T cell activation. Thus, IL-17 plays an important role in activating T cells in allergen-specific T cell-mediated immune responses.

The pathologic roles of these cytokines and the functional interrelationship among these cytokines in the development of arthritis, aortitis, and dermatitis that develop in IL-1Ra$^{-/-}$ mice, however, remain to be elucidated. In this paper, we will describe the roles for TNFα and IL-17 in the development of diseases resulting from excess IL-1 signaling.

8.2 The Roles of TNFα and IL-17 in the Development of Arthritis

8.2.1 Development of Autoimmune Arthritis in IL-1Ra$^{-/-}$ Mice

IL-1Ra$^{-/-}$ mice on the BALB/c background developed arthritis spontaneously; arthritis began to develop at 5 weeks of age and almost all of the animals suffered from arthritis at 12 weeks of age (Horai et al. 2000). The incidence of arthritis in IL-1Ra$^{-/-}$ mice differed among different genetic backgrounds; the incidence was high on the BALB/c background, but low on the C57BL/6 background, suggesting involvement of BALB/c-specific host genes. The histopathology of the lesions demonstrated a marked synovial and periarticular inflammation with articular erosion caused by invasion of granulation tissues, closely resembling the phenotype of RA in humans (Fig. 1). Osteoclast activation was obvious at the pannus, while inflammatory cell infiltration, consisting mostly of neutrophils, and fibrin clots were detectable in the synovial spaces.

Both total immunoglobulin levels and the levels of autoantibodies specific for immunoglobulin, type II collagen, and dsDNA were elevated in IL-1Ra$^{-/-}$ mice (Horai et al. 2000). The development of arthritis was completely suppressed in *scid/scid*-IL-1Ra$^{-/-}$ mice and adoptive transfer of IL-1Ra$^{-/-}$ T cells induced arthritis in *nu/nu* mice, suggesting a critical role for T cells in the pathogenesis of arthritis in this animal model (Horai et al. 2004). Arthritogenic activated and/or memory T cells were generated in IL-1Ra$^{-/-}$ mice, because T cells from arthritic IL-1Ra$^{-/-}$ mice could transfer disease more efficiently than those from nonarthritic mice. Since IL-1Ra mRNA expression was observed even in unstimulated T cells at low levels and the expression was enhanced

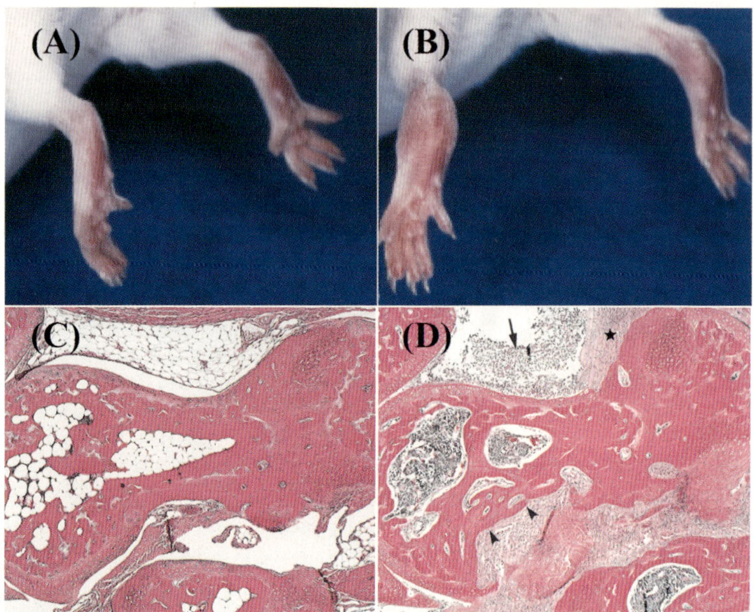

Fig. 1. Histopathology of ankle joints from BALB/c-IL-1Ra$^{-/-}$ mice. The ankles of a normal IL-1Ra$^{+/+}$ mouse (**A**) and an affected IL-1Ra$^{-/-}$ mouse (**B**) were examined at 16 weeks of age. Swelling and redness of the joints were observed in the IL-1Ra$^{-/-}$ mouse. Microscopic observation of the joints of IL-1Ra$^{+/+}$ (**C**) and IL-1Ra$^{-/-}$ (**D**) mice demonstrated the erosive destruction of bone in the IL-1Ra$^{-/-}$ mouse (*arrowheads*). The infiltration of inflammatory cells (*arrows*) and the proliferation of the synovial membrane lining cells (*asterisk*) were remarkable

in activated T cells, these data suggest that, although IL-1Ra is produced by cells of various types including monocytes and macrophages in the synovial lining layer, T cell-derived IL-1Ra likely regulates T cell activity in an autocrine manner.

We have previously shown that antigen-presenting cell (APC)-derived IL-1 can promote T cell activation through the induction of CD40L and OX40 on T cells, suggesting that IL-1 is an important regulator of acquired immune responses (Nakae et al. 2001b). In support of this notion, antibody production against sheep red blood cells increased in

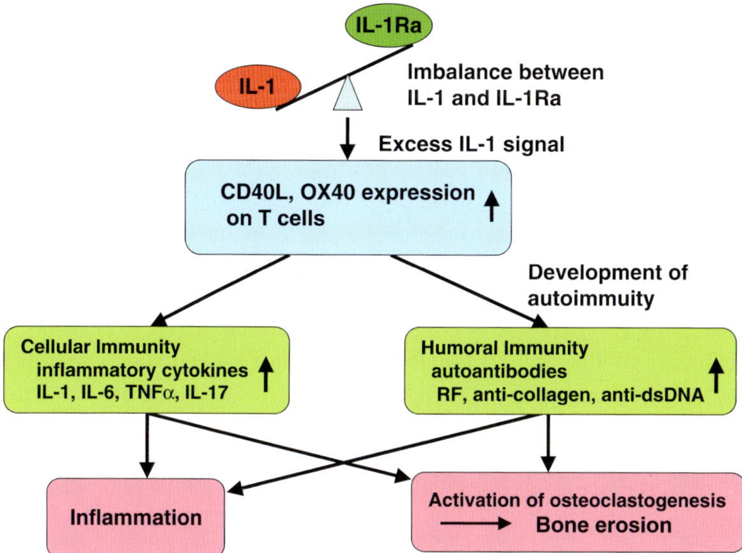

Fig. 2. Excess IL-1 signaling causes autoimmune arthritis

IL-1Ra$^{-/-}$ mice and decreased in IL-1$^{-/-}$ mice (Nakae et al. 2001a). Thus, it is suggested that, in the absence of IL-1Ra, even physiological levels of IL-1, which is constitutively expressed in the joints, can activate naive T cells excessively, resulting in the development of autoimmunity. These observations indicate that the balance between IL-1 and IL-1Ra is critical for homeostasis of the immune system (Fig. 2).

8.2.2 The Role of TNFα in the Development of Arthritis

The expression of numerous inflammatory cytokines, including TNFα and IL-17, was augmented in the joints of IL-1Ra$^{-/-}$ mice. To elucidate the roles of these cytokines in arthritis development, we examined the effect of cytokine deficiency on disease initiation and progression. We determined that the development of arthritis was markedly suppressed in TNFα$^{-/-}$ mice, indicating the crucial role for TNFα in pathogenesis (Horai et al. 2004). A dominant role for TNFα in the development of

RA has been demonstrated by recent clinical trials using therapeutic anti-TNFα antibody (Feldmann and Maini 2001). Studies in the type II collagen-induced arthritis (CIA) mouse model also support this notion (Thorbecke et al. 1992; Joosten et al. 1999; Feldmann and Maini 2001).

While bone marrow (BM) cell transplantation from TNF$\alpha^{+/+}$-IL-1Ra$^{-/-}$ mice into γ-ray-irradiated wild-type (WT) recipient mice induced arthritis, TNF$\alpha^{-/-}$-IL-1Ra$^{-/-}$ BM cells could not induce arthritis. These results indicate that BM-derived cells are responsible for the production of TNFα (Horai et al. 2004). It is interesting, however, that TNF$\alpha^{-/-}$-IL-1Ra$^{-/-}$ BM cells could induce arthritis when inoculated into IL-1Ra$^{-/-}$, but not WT recipient mice. Since TNFα expression is augmented in the joints of IL-1Ra$^{-/-}$ mice, this TNFα may compensate for the deficiency in the BM-derived cells. T cells are sensitive to irradiation, while synovial lining cells are relatively resistant to irradiation. Some of the synovial lining cells, such as type A cells, are derived from the BM origin and are eventually replaced by donor cells after BM cell transplantation. In recipient IL-1Ra$^{-/-}$ mice, TNFα is likely produced by these synovial lining cells, which may donate the arthritogenic milieu observed in IL-1Ra$^{-/-}$ mice in BM cell transfer experiments.

Furthermore, we found that the transfer of TNF$\alpha^{-/-}$-IL-1Ra$^{-/-}$ T cells into *scid/scid* mice did not promote the development of arthritis as robustly as TNF$\alpha^{+/+}$-IL-1Ra$^{-/-}$ T cells, suggesting that T cells are not efficiently sensitized in TNF$\alpha^{-/-}$ mice (Table 1). Thus, these results

Table 1. Inefficiency of the development of arthritis in *scid/scid* mice transferred with T cells from TNF$\alpha^{-/-}$-IL-1Ra$^{-/-}$ mice

Donor	Arthritis	
	Incidence	Score
TNF$\alpha^{+/+}$-IL-1Ra$^{-/-}$	4/5	5.0
TNF$\alpha^{-/-}$-IL-1Ra$^{-/-}$	1/11*	7.5
TNF$\alpha^{+/+}$-IL-1Ra$^{+/+}$	0/5**	0.0

Splenic CD4$^+$T cells (5×10^7) were transferred into BALB.B-*scid/scid* mice, and the development of arthritis was inspected after 12 weeks
* $p < 0.05$ vs TNF$\alpha^{+/+}$-IL-1Ra$^{-/-}$ mice by χ^2 test
** $p < 0.01$ vs TNF$\alpha^{+/+}$-IL-1Ra$^{-/-}$ mice by χ^2 test

suggest that TNFα derived from both T cells and synovial lining cells is involved in the development of arthritis.

We previously showed that IL-1 plays an important role in the enhancement of T cell–APC interactions through the induction of CD40L and OX40 on T cells (Nakae et al. 2001b). Consistently with this observation, CD40L and OX40 expression were enhanced in T cells stimulated with antigen-bearing IL-1Ra$^{-/-}$ APCs in comparison to WT APCs. On the other hand, ligation of CD40 on APCs by CD40L induces OX40L expression and TNFα production by APCs (van Kooten and Banchereau 2000; Weinberg 2002). Furthermore, we have demonstrated that TNFα induces OX40 expression on T cells (Horai et al. 2004). Thus, upon interaction with antigens, APCs produce IL-1. IL-1, in turn, activates T cells, resulting in the induction of CD40L. The CD40L-CD40 interaction induces APCs to produce OX40L and TNFα, resulting in the induction of OX40 on T cells. Therefore, it is suggested that TNFα plays an important role in the sensitization of T cells and contributes to the development of autoimmunity.

It is known that TNFα$^{-/-}$ mice lack splenic primary B cell follicles and cannot form either organized follicular dendritic cell (DC) networks or germinal centers in the spleen and peripheral lymphatic organs (Pasparakis et al. 1996). Prolonged antibody responses are generally impaired in TNFα$^{-/-}$ mice, although Ig class-switching is not completely deficient. It is therefore reasonable to suppose that these functions of TNFα in the humoral immune responses may also contribute to the development of autoimmunity in IL-1Ra$^{-/-}$ mice. However, since we could not induce arthritis by transferring IL-1Ra$^{-/-}$ mouse serum into WT mice, only a weak contribution of humoral immune responses is suggested in this animal model (Horai and Nakajima, unpublished results).

On the other hand, it is known that TNFα elicits inflammation by activating and recruiting inflammatory cells and inducing proinflammatory cytokines and chemokines, such as IL-1, IL-6, and CXCLIO (Pang et al. 1994; Nakae et al. 2003b). In this context, mouse models that exhibit higher amounts of TNFα protein, such as transgenic (Tg) mice carrying the TNFα gene or mice deficient for the TNFα AU-rich element (TNF$^{\Delta ARE}$), spontaneously develop arthritis (Keffer et al. 1991; Kontoyiannis et al. 1999). It was suggested that innate and/or stromal mechanisms rather than T cell-mediated autoimmune mechanisms are involved in the development of arthritis, because arthritis develops in RAG1$^{-/-}$ background, although Crohn's-like disease is induced by an immune-mediated mechanism in the same model (Kontoyiannis et al. 1999). Likewise, inflammatory cytokines, such as IL-1 and TNFα, but not IL-6, play important roles in the effector phase of the disease in the K/BxN model, although the effect of TNFα deficiency was not as strong in this model compared to that seen in the IL-1Ra$^{-/-}$ mice (Ji et al. 2002). Taken together, these observations suggest that TNFα plays important roles in both sensitization of T cells and elicitation of inflammation in the development of arthritis in IL-1Ra$^{-/-}$ mice.

8.2.3   The Role of IL-17 in the Development of Arthritis

IL-17 levels in IL-1Ra$^{-/-}$ mouse joints were elevated from the levels seen in wild-type mice. After stimulation with CD3, IL-17 production was greatly enhanced in IL-1Ra$^{-/-}$ T cells (Nakae et al. 2003d). In our examination of the development of arthritis in IL-17$^{-/-}$ mice, we demonstrated that IL-17-deficiency completely suppressed the onset of disease in IL-1Ra$^{-/-}$ mice (Nakae et al. 2003d). Joint inflammation was also suppressed in IL-17$^{-/-}$-human T cell leukemia virus type I (HTLV-I) Tg mice carrying the HTLV-I *tax* gene, another RA model in which arthritis develops spontaneously (Iwakura et al. 1991; unpublished observation). An important role for IL-17 was also indicated in the CIA model (Nakae et al. 2003c). We have shown that, upon stimulation with ovalbumin (OVA), OVA-specific T cell proliferation was low in T cells from IL-17$^{-/-}$-DO11.10 mice (Nakae et al. 2002, 2003d), mice carrying an OVA-specific T cell receptor transgene. These results suggest that IL-17 is involved in T cell priming. Consistent with this notion, we demonstrated that the sensitization of T cells following immunization with type II collagen was significantly reduced in IL-17$^{-/-}$ mice (Nakae et al. 2003c). Nonetheless, since both the incidence and the severity score were reduced in IL-17$^{-/-}$ mice in CIA, IL-17 may function not only at the sensitization phase but also the elicitation phase.

As mentioned above, IL-1Ra production by T cells is critical in the regulation of T cell activity by acting on T cells in an autocrine manner

(Horai et al. 2004). It is known that IL-1 induces CD40L on T cells, and CD40 signaling activates TNFα expression in APCs (van Kooten and Bancbereau 2000; Nakae et al. 2001b). Since the TNFα induces OX40 expression on T cells (Horai et al. 2004) and IL-17 production by T cells was induced by OX40 activation (Nakae et al. 2003d), TNFα-mediated induction of OX40 expression in T cells may induce production of IL-17, resulting in the exacerbation of inflammation. Thus, it was suggested that both CD40L-CD40 and OX40L-OX40 play important roles in the development of autoimmunity. In agreement with this notion, blockade of the CD40L–CD40 or OX40–OX40L interaction inhibited arthritis development in IL-1Ra$^{-/-}$ mice (Horai et al. 2004). These observations suggest that T cell-dependent autoimmunity is induced in IL-1Ra$^{-/-}$ mice through the induction of TNFα and IL-17, as the downstream mediators of the IL-1 action, and these cytokines also play important roles in the elicitation phase of inflammation (Fig. 3).

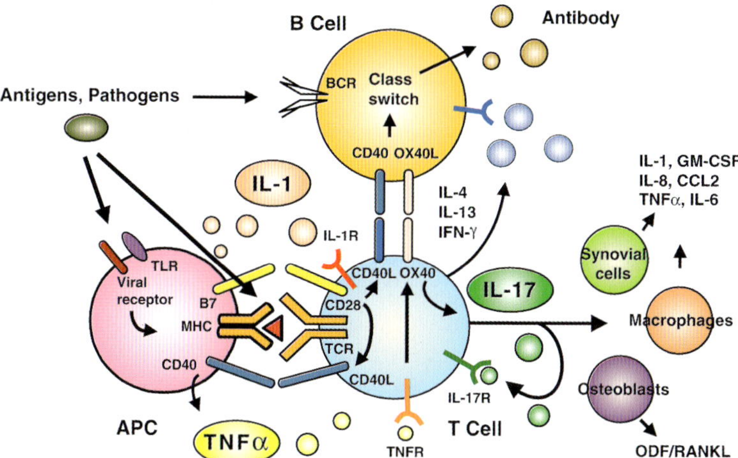

Fig. 3. Crucial roles for IL-17 and TNFα, downstream of IL-1 signaling, in the pathogenesis of arthritis

8.3　The Roles of TNFα and IL-17 in the Development of Aortitis

8.3.1　Development of Aortitis in IL-1Ra$^{-/-}$ Mice

IL-1Ra$^{-/-}$ mice on the BALB/c background spontaneously developed arterial inflammation at 4 weeks of age. Approximately 50% of the mice were affected by 12 weeks (Matsuki et al. 2005). A similar observation was reported in IL-1Ra$^{-/-}$ mice on the 129/Ola x MF1 background (Nicklin et al. 2000). On the C57BL/6 background, however, there were no signs of arterial inflammation, suggesting the significant involvement of background genes in the development of aortitis, a similar observation that has been made for arthritis (Horai et al. 2000). In F2 hybrids of BALB/c- and C57BL/6-IL-1Ra$^{-/-}$ mice, arthritis was rare but aortic inflammation was common, indicating that the sets of background modifier genes that cause susceptibility to each disease are not fully overlapping (Shepherd et al. 2004).

Inflammation of the cardiovascular system was observed preferentially at the aortic root of IL-1Ra$^{-/-}$ mice (Fig. 4) (Matsuki et al. 2005). The infiltration of monocytes and occasionally neutrophils was observed in the aorta and aortic valve. A loss of elastic lamellae in the aortic media could be observed by histological examinations. Monocytes/macrophages and some neutrophils had infiltrated the inflammatory sites within the aortic sinus. Thus, the aortic inflammation in these animals may have characteristics of both acute and chronic phases of disease. We identified numerous examples of neovascularization within severe lesions. Chondrocyte-like cells were observed in the majority of IL-1Ra$^{-/-}$ mouse aortas; no such cells could be observed in the aortas of WT mice. Calcification of the media of the aorta was observed in a subset of IL-1Ra$^{-/-}$ mice. As calcification of the media, involving the degradation of smooth muscle cells, is a sign of degenerative processes (Tanimura et al. 1986a, 1986b), this result suggests the involvement of an immune response in this pathology. These mice suffered from mild aortic stenosis and hyperplasia of the interventricular septum and left ventricular posterior walls. In agreement with previous reports, these pathological findings resemble aspects of Takayasu arteritis or polyarteritis nodosa in humans (Nicklin et al. 2000).

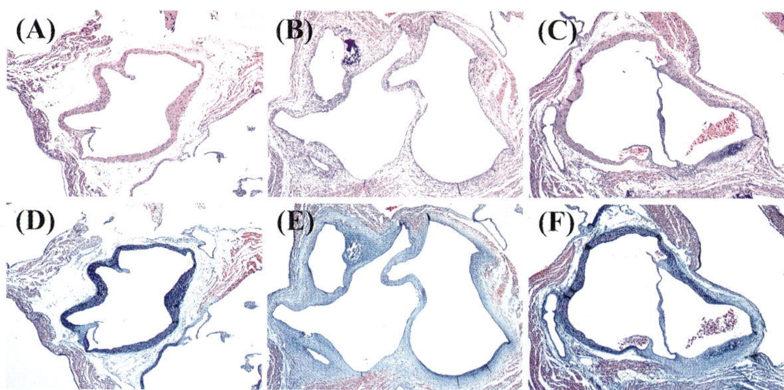

Fig. 4. Attenuation of the development of aortitis in IL-1Ra$^{-/-}$ mice by IL-17 deficiency. Sections of the aorta were examined by staining with (**A–C**) hematoxylin and eosin and (**D–F**) Masson's trichrome (Isoda et al. 2002). **A, D** Sections of the aortic valve (score 0) from a 20-week-old unaffected IL-17$^{+/-}$-IL-1Ra$^{-/-}$ mouse. **B, E** Sections from a 20-week-old affected IL-17$^{+/-}$-IL-1Ra$^{-/-}$ mouse. Severe inflammatory cell infiltration and loss of elastic lamellae over greater than two-thirds of the media of the aortic sinus are observed (score 3). **C, F** Sections from an 20-week-old IL-17$^{-/-}$-IL-1Ra$^{-/-}$ mice. Mild inflammatory cell infiltration and loss of elastic lamellae are observed (score 2)

Using peripheral T cell transplantation, we also examined the role of T cells in the development of aortitis (Matsuki et al. 2005). Purified T cells from the spleens and lymph nodes of 6- to 8-week-old IL-1Ra$^{-/-}$ mice were transplanted into BALB/c-*nu/nu* mice, and the development of aortitis in the recipient mice was analyzed 10 weeks after transplantation. Twelve of the 13 recipient mice developed aortitis, indicating that T cells are crucial in the pathogenesis of aortitis. As arthritis is also induced by IL-1Ra$^{-/-}$ T cell transplantation, the pathogenesis of aortitis likely utilizes a similar mechanism as that seen in arthritis, in which T cell-mediated autoimmunity caused by excess IL-1 signaling is involved.

8.3.2　The Roles of TNFα and IL-17 in the Development of Aortitis

We examined the roles of TNFα and IL-17 in the development of aortitis by intercrossing these cytokine-deficient mice to IL-1Ra$^{-/-}$ mice. The aortic valves of these cytokine-deficient IL-1Ra$^{-/-}$ mice were analyzed histologically. Interestingly, TNFα$^{-/-}$-IL-1Ra$^{-/-}$ mice showed no signs of arterial inflammation at 8 and 14 weeks of age, while approximately 50% of the TNFα$^{+/+}$-IL-1Ra$^{-/-}$ mice developed aortitis at these ages (Matsuki et al. 2005). The incidence of aortitis in IL-17$^{-/-}$-IL-1Ra$^{-/-}$ mice was similar to IL-17$^{+/-}$-IL-1Ra$^{-/-}$ mice at 20–28 weeks of age (Table 2). The disease severity score, however, was significantly reduced in these IL17$^{-/-}$-IL-1Ra$^{-/-}$ mice (Fig. 4). Thus, in IL-1Ra$^{-/-}$ mice, TNFα is crucial for the development of aortitis. While IL-17 is not essential for aortitis development, it aggravates the disease, appearing to function at both the elicitation of inflammation and the sensitization of T cells.

As mentioned already, we have demonstrated that T cell-derived TNFα plays an important role for the sensitization of T cells in the development of autoimmunity in IL-1Ra$^{-/-}$ mice (Horai et al. 2004). Other investigators have also reported the production of TNFα in T cells (Ramshaw et al. 1994; Sakaguchi et al. 1995) and the presence of TNF receptors in aortic smooth muscle and endothelial cells (Field et al. 1997). Thus, upon T cell activation, T cells produce TNFα, and this T-cell derived TNFα may activate endothelial cells to produce various

Table 2. Suppression of the development of aortitis in IL-17$^{-/-}$-IL-1Ra$^{-/-}$ mice

Genotype	Incidence (%) Severity score 20, 28 weeks
IL-17$^{+/-}$-IL-1Ra$^{-/-}$	5/6 (83%) 2.8
IL-17$^{-/-}$-IL-1Ra$^{-/-}$	6/13 (46%) 1.8*

* $p < 0.05$ vs IL-17$^{+/-}$-IL-1Ra$^{-/-}$ mice by Mann-Whitney U test

inflammatory cytokines and chemokines, resulting in the development of inflammation (Kollias and Kontoyiannis 2002). It is also known that TNFα induces the expression of vascular cell adhesion molecule-1 in endothelial cells; this promotes the early adhesion of mononuclear leukocytes to the arterial endothelium at sites of inflammation (Feldmann 2002).

Consistent with our observations, it was recently reported that Infliximab, an anti-TNFα antibody, improved endothelial dysfunction in antineutrophil cytoplasmic antibody-associated systemic vasculitis in humans (Booth et al. 2004). Although the etiopathogenesis of this vasculitis has not been completely elucidated, it is thought that both aortitis in IL-1Ra$^{-/-}$ mice and antineutrophil cytoplasmic antibody-associated systemic vasculitis in humans share a similar pathogenic process involving TNFα. These observations provide new insight into the pathogenesis of vasculitis, and the IL-1Ra$^{-/-}$ mice should be a useful model for the study of the pathogenic mechanisms of vasculitis (Fig. 5).

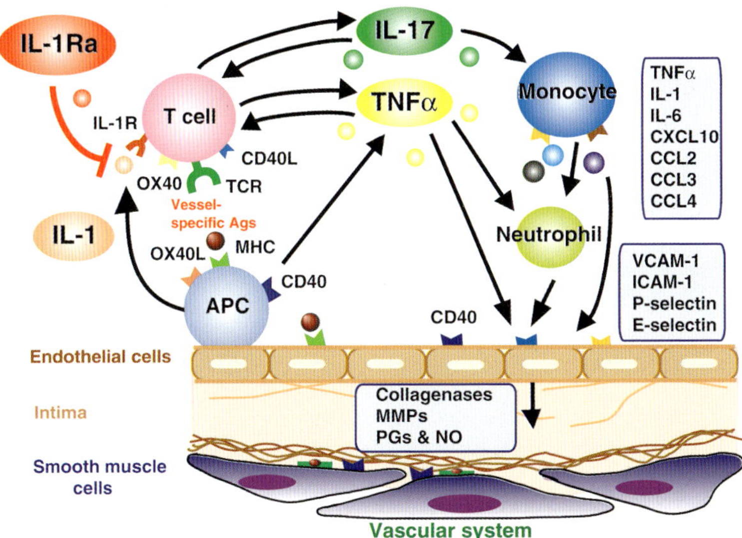

Fig. 5. Pathogenesis of aortitis: TNFα plays an important role in the development of aortitis in IL-1Ra$^{-/-}$ mice

8.4   The Role of TNFα in the Development of Dermatitis

8.4.1   Development of Psoriasis-Like Dermatitis in IL-1Ra$^{-/-}$ Mice

Psoriasis-like skin disease, first reported by Shepherd et al. (2004), was evident in IL-1Ra$^{-/-}$ mice on the BALB/c background. However, we could not identify disease in animals on the C57BL/6 background, indicating that strain-specific background genes are also involved in the development of this disease. The mice on the BALB/c background develop redness and scaling of their ears and tail (Fig. 6), a pathology characterized by extensive thickening of the epidermis associated with hyperkeratosis of the skin. The majority of keratinocytes retained their nucleus in the cornified cellular layer. We also observed massive neutrophil infiltration into the epidermis and dermis. With increasing disease progression, the epidermal layer gradually penetrated into the dermal layer, forming ridges. Aseptic microabscesses formed under the skin. CD4$^+$ T cells were occasionally observed within the dermis.

Interestingly, however, significant disease developed in *scid/scid*-IL-1Ra$^{-/-}$ mice. Furthermore, IL-1Ra$^{-/-}$ T cell transfer could not induce dermatitis in *nu/nu* mice, in contrast to cases of arthritis or aortitis in

(A)   **(B)**

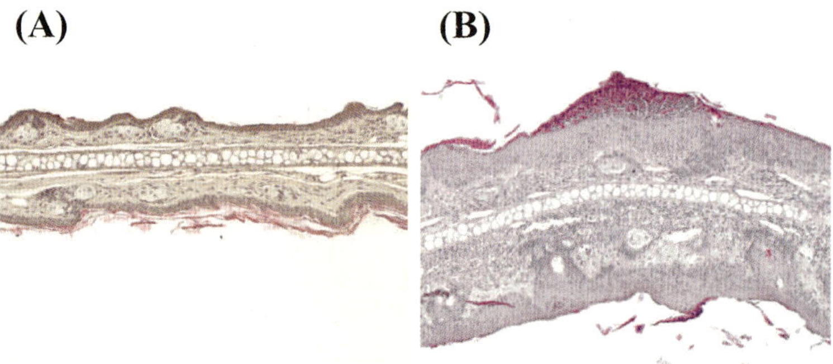

Fig. 6. Histological examination of the skin in IL-1Ra$^{-/-}$ mice. **A** Hematoxylin and eosin staining of the ear pinna of a normal 20-week-old WT mouse. **B** An affected IL-1Ra$^{-/-}$ mouse at 20 weeks of age. The epidermis becomes thickened and hypertrophic associated with hyperkeratosis of the skin. Massive infiltration of neutrophils into the epidermis and dermis are observed in IL-1Ra$^{-/-}$ mouse

which diseases could be induced by T cell transfer. Thus, in this case, an autoimmune process is not likely to be involved in disease pathogenesis; rather, excess IL-1 signaling directly induces inflammation within the skin.

In humans, the involvement of IL-1 in the development of psoriasis has not been elucidated completely. Some studies have indicated that IL-1α concentrations were reduced in psoriatic lesional skin as compared to nonlesional and normal skin, although IL-1β concentrations were increased (Cooper et al. 1990; Debets et al. 1997). Our data clearly show that excess IL-1 signaling can induce psoriasis-like lesions in mice, suggesting involvement of IL-1 in the development of psoriasis in humans. In agreement with our observations, Tg mice that express IL-1α from K14 promoter in the basal epidermis also develop scaly and erythematous inflammatory skin lesions (Groves et al. 1995).

8.4.2 The Role of TNFα in the Development of Dermatitis

The development of dermatitis in IL-1Ra$^{-/-}$ mice was completely absent in TNFα-deficient IL-1Ra$^{-/-}$ mice, indicating a crucial role for TNFα in disease pathogenesis. The importance of IL-1 and TNFα were also seen in contart hypersensitivity (CHS) reactions, in which antigen-specific CD4$^+$ T cells play a central role. 2, 4, 6-trinitrochlorobenzene (TNCB)-induced CHS was suppressed in IL-1$\alpha/\beta^{-/-}$ and IL-1$\alpha^{-/-}$, but not IL-1$\beta^{-/-}$, mice, and these responses were augmented in IL-1Ra$^{-/-}$ mice, suggesting an important role for IL-1α in CHS responses (Nakae et al. 2001c, 2003b). We demonstrated that the IL-1 produced by APCs of the epidermis enhances the sensitization of allergen-specific T cells and induces inflammation via TNFα production during the elicitation phase (Nakae et al. 2003b). TNFα elicits inflammatory cell infiltration in the skin through the induction of CXCLIO.

TNFα production is increased in psoriatic lesional skin as compared to nonlesional and healthy skin (Ettehadi et al. 1994). Moreover, direct correlation between TNFα concentration either at the lesional skin or serum levels and the psoriasis area severity index scores has been reported (Bonifati et al. 1994). Thus, TNFα may also be involved in the development of psoriasis in humans (Fig. 7).

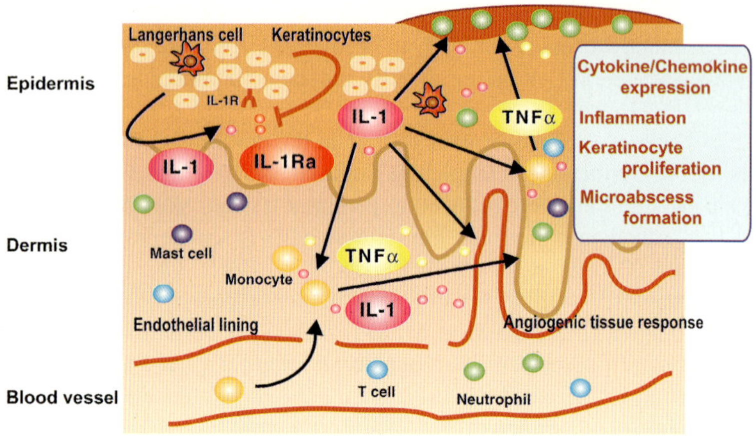

Fig. 7. Pathogenesis of dermatitis

In human psoriasis, the importance of an acquired immune response is suggested, because linkage association with MHC class I is observed (Elder et al. 1994), the dermis and the epidermis are heavily infiltrated by CD4[+] T cells (Baker and Fry 1992), cyclosporin A efficiently suppresses psoriasis (Griffiths et al. 1989), and depletion of CD25[+] T cells ameliorates the disease (Gottlieb et al. 1995). Furthermore, it was shown that injection of prepsoriatic skin engrafted onto *scid/scid* mice with CD4[+] T cells induces psoriasis (Nickoloff and Wrone-Smith 1999). However, so far, no conclusive evidence has been presented for the involvement of autoimmunity in this disease. On the other hand, it has been argued that keratinocytes of psoriatics suffer from an intrinsic abnormality in the regulation of their activation by cytokines, which trigger proliferation and migration, and stimulated keratinocytes may act as initiators of an inflammatory process by means of the secretion of various cytokines able to induce the expression of cell adhesion molecules and the recruitment of inflammatory cells (Bonifati and Ameglio 1999; Shepherd et al. 2004). Our observations suggest that keratinocyte-derived pathogenesis rather than an autoimmune mechanism is involved in the development of dermatitis in IL-1Ra$^{-/-}$ mice. Since the infiltration of inflammatory cells is not prominent at the beginning of the disease and gradually increases, immune mechanisms may be involved at the later phase.

8.5 Conclusion

We have demonstrated that a variety of inflammatory diseases, including arthritis, aortitis, and psoriatic dermatitis, develop spontaneously in IL-1Ra$^{-/-}$ mice. Although excess IL-1 signaling is responsible for the development of these diseases, the pathogenic mechanisms differ significantly; both arthritis and aortitis result from the development of autoimmunity, while such autoimmune processes are not involved in the development of dermatitis. Both TNFα and IL-17 play important roles in the activation of T cells downstream of IL-1 signaling, in addition to the roles in the elicitation of inflammation. TNFα activates T cells by inducing OX40 expression, leading to increased IL-17 production. Although IL-17 was also shown to be involved in the sensitization of T cells, the mechanism underlying this activation remains to be elucidated. Thus, both cytokines play crucial roles in the development of the autoimmunity that can cause arthritis and aortitis in this knockout mouse model. In contrast, in the case of psoriatic dermatitis, excess IL-1 signaling and TNFα signaling directly induce inflammation of the skin without the involvement of autoimmunity. Thus, IL-1 and TNFα have dual functions, the activation of T cells and the direct induction of inflammation. It is interesting that excess IL-1signaling can induce several different diseases in an animal via different mechanisms. In any case, these observations suggest that the suppression of IL-1, TNFα, and IL-17 is important in the control of inflammatory diseases; suppression of cytokine expression or action should be beneficial for the treatment of these diseases.

Acknowledgements. This work was supported by grants from the Ministry of Education, Culture, Sports, Science and Technology of Japan, and The Ministry of Health, Labour and Welfare of Japan.

References

Aarvak T, Chabaud M, Miossec P, Natvig JB (1999) IL-17 is produced by some proinflammatory Th1/Th0 cells but not by Th2 cells. J Immunol 162:1246–1251

Aggarwal S, Gurney AL (2002) IL-17: prototype member of an emerging cytokine family. J Leukoc Biol 71:1–8

Albanesi C, Scarponi C, Cavani A, Federici M, Nasorri F, Girolomoni G (2000) Interleukin-17 is produced by both Th1 and Th2 lymphocytes, and modulates interferon-gamma- and interleukin-4-induced activation of human keratinocytes. J Invest Dermatol 115:81–87

Baker BS, Fry L (1992) The immunology of psoriasis. Br J Dermatol 126:1–9

Bonifati C, Ameglio F (1999) Cytokines in psoriasis. Int J Dermatol 38:241–251

Bonifati C, Carducci M, Cordiali Fei P, Trento E, Sacerdoti G, Fazio M, Ameglio F (1994) Correlated increases of tumour necrosis factor-alpha, interleukin-6 and granulocyte monocyte-colony stimulating factor levels in suction blister fluids and sera of psoriatic patients – relationships with disease severity. Clin Exp Dermatol 19:383–387

Booth AD, Jayne DR, Kharbanda RK, McEniery CM, Mackenzie IS, Brown J, Wilkinson IB (2004) Infliximab improves endothelial dysfunction in systemic vasculitis: a model of vascular inflammation. Circulation 109:1718–1723

Carswell EA, Old LJ, Kassel RL, Green S, Fiore N, Williamson B (1975) An endotoxin-induced serum factor that causes necrosis of tumors. Proc Natl Acad Sci U S A 72:3666–3670

Carter DB, Deibel MR Jr, Dunn CJ, Tomich C-SC, Laborde AL, Slightom JL, Berger AE, Bienkowski MJ, Sun FF, McEwan RN, Harris PK, Yem AW, Waszak GA, Chosay JG, Sieu LC, Hardee MM, Zurcher-Neely HA, Reardon IM, Heinrikson RL, Truesdell SE, Shelly JA, Eessalu TE, Taylor BM, Tracey DE (1990) Purification, cloning, expression and biological characterization of an interleukin-1 receptor antagonist protein. Nature 344:633–638

Colotta F, Re F, Muzio M, Bertini R, Polentarutti N, Sironi M, Giri JG, Dower SK, Sims JE, Mantovani A (1993) Interleukin-1 type II receptor: a decoy target for IL-1 that is regulated by IL-4. Science 261:472–475

Cooper KD, Hammerberg C, Baadsgaard O, Elder JT, Chan LS, Sauder DN, Voorhees JJ, Fisher G (1990) IL-1 activity is reduced in psoriatic skin. Decreased IL-1 alpha and increased nonfunctional IL-1 beta. J Immunol 144:4593–4603

Davis P, MacIntyre DE (1992) Prostaglandins and inflammation. In: Snyderman R (ed) Inflammation: basic principles and clinical correlates. Raven Press, New York, pp 123–137

Debets R, Hegmans JP, Croughs P, Troost RJ, Prins JB, Benner R, Prens EP (1997) The IL-1 system in psoriatic skin: IL-1 antagonist sphere of influence in lesional psoriatic epidermis. J Immunol 158:2955–2963

Dinarello CA (1991) Interleukin-1 and interleukin-1 antagonism. Blood 77:1627–1652

Dinarello CA (1996) Biologic basis for interleukin-1 in disease. Blood 87:2095–2147

Durum SK, Oppenheim JJ (1993) Proinflammatory cytokines and immunity. In: Paul WE (ed) Fundamental immunology, 3rd edn. Raven Press, New York, pp 801–835

Elder JT, Nair RP, Guo SW, Henseler T, Christophers E, Voorhees JJ (1994) The genetics of psoriasis. Arch Dermatol 130:216–224

Ettehadi P, Greaves MW, Wallach D, Aderka D, Camp RD (1994) Elevated tumour necrosis factor-alpha (TNF-alpha) biological activity in psoriatic skin lesions. Clin Exp Immunol 96:146–151

Feldmann M (2002) Development of anti-TNF therapy for rheumatoid arthritis. Nat Rev Immunol 2:364–371

Feldmann M, Maini RN (2001) Anti-TNF alpha therapy of rheumatoid arthritis: what have we learned? Annu Rev Immunol 19:163–196

Feldmann M, Brennan FM, Maini RN (1996) Role of cytokines in rheumatoid arthritis. Annu Rev Immunol 14:397–440

Ferretti S, Bonneau O, Dubois GR, Jones CE, Trifilieff A (2003) IL-17, produced by lymphocytes and neutrophils, is necessary for lipopolysaccharide-induced airway neutrophilia: IL-15 as a possible trigger. J Immunol 170:2106–2112

Field M, Cook A, Gallagher G (1997) Immuno-localisation of tumour necrosis factor and its receptors in temporal arteritis. Rheumatol Int 17:113–118

Fowlkes JL, Winkler MK (2002) Exploring the interface between metallo-proteinase activity and growth factor and cytokine bioavailability. Cytokine Growth Factor Rev 13:277–287

Gottlieb SL, Gilleaudeau P, Johnson R, Estes L, Woodworth TG, Gottlieb AB, Krueger JG (1995) Response of psoriasis to a lymphocyte-selective toxin (DAB389IL-2) suggests a primary immune, but not keratinocyte, pathogenic basis. Nat Med 1:442–447

Greenfeder SA, Nunes P, Kwee L, Labow M, Chizzonite RA, Ju G (1995) Molecular cloning and characterization of a second subunit of the interleukin 1 receptor complex. J Biol Chem 270:13757–13765

Griffiths CE, Powles AV, McFadden J, Baker BS, Valdimarsson H, Fry L (1989) Long-term cyclosporin for psoriasis. Br J Dermatol 120:253–260

Groves RW, Mizutani H, Kieffer JD, Kupper TS (1995) Inflammatory skin disease in transgenic mice that express high levels of interleukin 1 alpha in basal epidermis. Proc Natl Acad Sci U S A 92:11874–11878

Hannum CH, Wilcox CJ, Arend WP, Joslin FG, Dripps DJ, Heimdal PL, Armes LG, Sommer A, Eisenberg SP, Thompson RC (1990) Interleukin-1 receptor antagonist activity of a human interleukin-1 inhibitor. Nature 343:336–340

Hirsch E, Irikura VM, Paul SM, Hirsh D (1996) Functions of interleukin 1 receptor antagonist in gene knockout and overproducing mice. Proc Natl Acad Sci U S A 93:11008–11013

Horai R, Saijo S, Tanioka H, Nakae S, Sudo K, Okahara A, Ikuse T, Asano M, Iwakura Y (2000) Development of chronic inflammatory arthropathy resembling rheumatoid arthritis in interleukin 1 receptor antagonist-deficient mice. J Exp Med 191:313–320

Horai R, Nakajima A, Habiro K, Kotani M, Nakae S, Matsuki T, Nambu A, Saijo S, Kotaki H, Sudo K, Okahara A, Tanioka H, Ikuse T, Ishii N, Schwartzberg PL, Abe R, Iwakura Y (2004) TNF-alpha is crucial for the development of autoimmune arthritis in IL-1 receptor antagonist-deficient mice. J Clin Invest 114:1603–1611.

Infante-Duarte C, Horton HF, Byrne MC, Kamradt T (2000) Microbial lipopeptides induce the production of IL-17 in Th cells. J Immunol 165:6107–6115

Isoda K, Nishikawa K, Kamezawa Y, Yoshida M, Kusuhara M, Moroi M, Tada N, Ohsuzu F (2002) Osteopontin plays an important role in the development of medial thickening and neointimal formation. Circ Res 91:77–82

Iwakura Y, Tosu M, Yoshida E, Takiguchi M, Sato K, Kitajima I, Nishioka K, Yamamoto H, Takeda T, Hatanaka M, Yamamoto H, Sekiguchi T (1991) Induction of inflammatory arthropathy resembling rheumatoid arthritis in mice transgenic for HTLV-I. Science 253:1026–1028

Ji H, Pettit A, Ohmura K, Ortiz-Lopez A, Duchatelle V, Degott C, Gravallese E, Mathis D, Benoist C (2002) Critical roles for interleukin 1 and tumor necrosis factor alpha in antibody-induced arthritis. J Exp Med 196:77–85

Joosten LA, Helsen MM, Saxne T, van De Loo FA, Heinegard D, van Den Berg WB (1999) IL-1 alpha beta blockade prevents cartilage and bone destruction in murine type II collagen-induced arthritis, whereas TNF-alpha blockade only ameliorates joint inflammation. J Immunol 163:5049–5055

Keffer J, Probert L, Cazlaris H, Georgopoulos S, Kaslaris E, Kioussis D, Kollias G (1991) Transgenic mice expressing human tumour necrosis factor: a predictive genetic model of arthritis. EMBO J 10:4025–4031

Kennedy J, Rossi DL, Zurawski SM, Vega F Jr, Kastelein RA, Wagner JL, Hannum CH, Zlotnik A (1996) Mouse IL-17: a cytokine preferentially expressed by alpha beta TCR + CD4$^-$CD8$^-$ T cells. J Interferon Cytokine Res 16:611–617

Kollias G, Kontoyiannis D (2002) Role of TNF/TNFR in autoimmunity: specific TNF receptor blockade may be advantageous to anti-TNF treatments. Cytokine Growth Factor Rev 13:315–321

Kolls JK, Linden A (2004) Interleukin-17 family members and inflammation. Immunity 21:467–476

Kontoyiannis D, Pasparakis M, Pizarro TT, Cominelli F, Kollias G (1999) Impaired on/off regulation of TNF biosynthesis in mice lacking TNF AU-rich elements: implications for joint and gut-associated immunopathologies. Immunity 10:387–398

Matsuki T, Isoda K, Horai R, Nakajima A, Aizawa Y, Suzuki K, Ohsuzu F, Iwakura Y (2005) Involvement of TNFα in the development of T cell-dependent aortitis in IL-1 receptor antagonist-deficient mice. Circulation (in press)

Moseley TA, Haudenschild DR, Rose L, Reddi AH (2003) Interleukin-17 family and IL-17 receptors. Cytokine Growth Factor Rev 14:155–174

Muzio M, Polentarutti N, Sironi M, Poli G, De Gioia L, Introna M, Mantovani A, Colotta F (1995) Cloning and characterization of a new isoform of the interleukin 1 receptor antagonist. J Exp Med 182:623–628

Nakae S, Asano M, Horai R, Iwakura Y (2001a) Interleukin-1 beta, but not interleukin-1 alpha, is required for T-cell-dependent antibody production. Immunology 104:402–409

Nakae S, Asano M, Horai R, Sakaguchi N, Iwakura Y (2001b) IL-1 enhances T cell-dependent antibody production through induction of CD40 ligand and OX40 on T cells. J Immunol 167:90–97

Nakae S, Naruse-Nakajima C, Sudo K, Horai R, Asano M, Iwakura Y (2001c) IL-1 alpha, but not IL-1 beta, is required for contact-allergen-specific T cell activation during the sensitization phase in contact hypersensitivity. Int Immunol 13:1471–1478

Nakae S, Komiyama Y, Nambu A, Sudo K, Iwase M, Homma I, Sekikawa K, Asano M, Iwakura Y (2002) Antigen-specific T cell sensitization is impaired in IL-17-deficient mice, causing suppression of allergic cellular and humoral responses. Immunity 17:375–387

Nakae S, Horai R, Komiyama Y, Nambu A, Asano M, Nakane A, Iwakura Y (2003a) The role of IL-1 in the immune system. In: Fantuzzi G (ed) Cytokine knockouts. Humana Press, Totowa, pp 99–109

Nakae S, Komiyama Y, Narumi S, Sudo K, Horai R, Tagawa Y, Sekikawa K, Matsushima K, Asano M, Iwakura Y (2003b) IL-1-induced tumor necrosis factor-alpha elicits inflammatory cell infiltration in the skin by inducing IFN-gamma-inducible protein 10 in the elicitation phase of the contact hypersensitivity response. Int Immunol 15:251–260

Nakae S, Nambu A, Sudo K, Iwakura Y (2003c) Suppression of immune induction of collagen-induced arthritis in IL-17-deficient mice. J Immunol 171:6173–6177

Nakae S, Saijo S, Horai R, Sudo K, Mori S, Iwakura Y (2003d) IL-17 production from activated T cells is required for the spontaneous development of destructive arthritis in mice deficient in IL-1 receptor antagonist. Proc Natl Acad Sci U S A 100:5986–5990

Nicklin MJ, Hughes DE, Barton JL, Ure JM, Duff GW (2000) Arterial inflammation in mice lacking the interleukin 1 receptor antagonist gene. J Exp Med 191:303–312

Nickoloff BJ, Wrone-Smith T (1999) Injection of pre-psoriatic skin with CD4[+] T cells induces psoriasis. Am J Pathol 155:145–158

Pang G, Couch L, Batey R, Clancy R, Cripps A (1994) GM-CSF, IL-1 alpha, IL-1 beta, IL-6, IL-8, IL-10, ICAM-1 and VCAM-1 gene expression and cytokine production in human duodenal fibroblasts stimulated with lipopolysaccharide, IL-1 alpha and TNF-alpha. Clin Exp Immunol 96:437–443

Pasparakis M, Alexopoulou L, Episkopou V, Kollias G (1996) Immune and inflammatory responses in TNF alpha-deficient mice: a critical requirement for TNF alpha in the formation of primary B cell follicles, follicular dendritic cell networks and germinal centers, and in the maturation of the humoral immune response. J Exp Med 184:1397–1411

Ramshaw AL, Roskell DE, Parums DV (1994) Cytokine gene expression in aortic adventitial inflammation associated with advanced atherosclerosis (chronic periaortitis). J Clin Pathol 47:721–727

Rouvier E, Luciani MF, Mattei MG, Denizot F, Golstein P (1993) CTLA-8, cloned from an activated T cell, bearing AU-rich messenger RNA instability sequences, and homologous to a herpesvirus saimiri gene. J Immunol 150:5445–5456

Sakaguchi M, Kato H, Nishiyori A, Sagawa K, Itoh K (1995) Characterization of CD4[+] T helper cells in patients with Kawasaki disease (KD): preferential production of tumour necrosis factor-alpha (TNF-alpha) by V beta 2– or V beta 8– CD4[+] T helper cells. Clin Exp Immunol 99:276–282

Shepherd J, Little MC, Nicklin MJ (2004) Psoriasis-like cutaneous inflammation in mice lacking interleukin-1 receptor antagonist. J Invest Dermatol 122:665–669

Shin HC, Benbernou N, Esnault S, Guenounou M (1999) Expression of IL-17 in human memory CD45RO+ T lymphocytes and its regulation by protein kinase A pathway. Cytokine 11:257–266

Sims JE, Gayle MA, Slack JL, Alderson MR, Bird TA, Giri JG, Colotta F, Re F, Mantovani A, Shanebeck K, Grabstein KH, Dower SK (1993) Interleukin 1 signaling occurs exclusively via the type I receptor. Proc Natl Acad Sci U S A 90:6155–6159

Tanimura A, McGregor DH, Anderson HC (1986a) Calcification in atherosclerosis. I. Human studies. J Exp Pathol 2:261–273

Tanimura A, McGregor DH, Anderson HC (1986b) Calcification in atherosclerosis. II. Animal studies. J Exp Pathol 2:275–297

Thorbecke GJ, Shah R, Leu CH, Kuruvilla AP, Hardison AM, Palladino MA (1992) Involvement of endogenous tumor necrosis factor alpha and transforming growth factor beta during induction of collagen type II arthritis in mice. Proc Natl Acad Sci U S A 89:7375–7379

Tocci MJ, Schmidt JA (1997) Interleukin-1: structure and function. In: Remick DG, Friedland JS (eds) Cytokines in health and disease, 2nd edn. Marcel Dekker, New York, pp 1–27

Van Kooten C, Banchereau J (2000) CD40-CD40 ligand. J Leukoc Biol 67:2–17

Weinberg AD (2002) OX40: Targeted immunotherapy – implications for tempering autoimmunity and enhancing vaccines. Trends Immunol 23:102–109

Werman A, Werman-Venkert R, White R, Lee JK, Werman B, Krelin Y, Voronov E, Dinarello CA, Apte RN (2004) The precursor form of IL-1α is an intracrine proinflammatory activator of transcription. Proc Nat Acad Sci U S A 101:2434–2439

Yao Z, Fanslow WC, Seldin MF, Rousseau AM, Painter SL, Comeau MR, Cohen JI, Spriggs MK (1995a) Herpesvirus Saimiri encodes a new cytokine, IL-17, which binds to a novel cytokine receptor. Immunity 3:811–821

Yao Z, Painter SL, Fanslow WC, Ulrich D, Macduff BM, Spriggs MK, Armitage RJ (1995b) Human IL-17: a novel cytokine derived from T cells. J Immunol 155:5483–5486

Ye P, Rodriguez FH, Kanaly S, Stocking KL, Schurr J, Schwarzenberger P, Oliver P, Huang W, Zhang P, Zhang J, Shellito JE, Bagby GJ, Nelson S, Charrier K, Peschon JJ, Kolls JK (2001) Requirement of interleukin 17 receptor signaling for lung CXC chemokine and granulocyte colony-stimulating factor expression, neutrophil recruitment, and host defense. J Exp Med 194:519–527

9 TGFβ-Mediated Immunoregulation

Y. Peng, L. Gorelik, Y. Laouar, M.O. Li, R.A. Flavell

9.1   Introduction 156
9.2   Materials and Methods 156
9.2.1   Transgenic Mice 156
9.2.2   Tumor Model 156
9.2.3   Infection Model 157
9.2.4   Diabetes Model 157
9.3   Results 157
9.3.1   TGFβ and Effector T Cells 157
9.3.2   TGFβ and Regulatory T Cells 158
9.4   Conclusions 159
References 160

Abstract. T cell homeostasis is required for normal immune responses and prevention of pathological responses. Transforming growth factor β (TGFβ) plays an essential role in that regulation. Owing to its broad expression and inhibitory effects on multiple immune cell types, TGFβ regulation is complex. Through recent advances in cell-specific targeting of TGFβ signaling in vivo, the role of TGFβ in T cell regulation is emerging. We demonstrated here a critical role for TGFβ in regulating effector vs regulatory T cell homeostasis.

9.1 Introduction

Understanding immune regulation and tolerance remains a major challenge in immunology. As an immunosuppressive cytokine, TGFβ plays a key role in attenuating excessive pathological immune responses. Mice deficient in TGFβ, developed severe multifocal inflammatory disease and died shortly after birth (Shull et al. 1992). Since TGFβ can be produced by and act on virtually all cell types, the regulatory network invoked by TGFβ remains incompletely understood. To investigate the role of TGFβ signaling in T cells, we engineered a strain of transgenic mice expressing a dominant negative TGFβ receptor II in T cells (Gorelik and Flavell 2000). In these mice, T cells were found to differentiate spontaneously into effector T cells, which led to the development of autoimmunity and enhanced tumor immunity and immune responses to pathogens. In addition, we took a gain-of-function approach to study TGFβ regulation of autoimmune diabetes by overexpressing this cytokine in the islets of the pancreas (Peng et al. 2004). We demonstrated that a short pulse of TGFβ expression could expand the regulatory T cell population and inhibit diabetes in NOD mice. These studies have thus identified multiple mechanisms by which TGFβ regulates T cell tolerance.

9.2 Materials and Methods

9.2.1 Transgenic Mice

Mice expressing a dominant negative mutant of TGFβ receptor II under the control of CD4 promoter (DNR mice) were described previously (Gorelik and Flavell 2000). All mice were kept under SPF conditions at Yale animal facility according to the approved protocols.

9.2.2 Tumor Model

The B16-F10 melanoma tumor cell line syngeneic to the C57BL/6 background was provided by A. Garen. EL4 cells were obtained from the American Type Culture Collection (Manassas, VA). Wild-type and DNR mice on C57BL/6 background were challenged with either B16-F10 cells i.v. or EL4 cells i.p. and monitored for tumor growth.

9.2.3   Infection Model

DNR mice were backcrossed onto the BALB/c background as described (Gorlik et al. 2002). Wild-type and DNR mice were infected in the right hind foot with *Leishmania major* promastigotes of the WR309 substrain and monitored for disease progression.

9.2.4   Diabetes Model

Under the control of a rat insulin promoter (RIP), tetracycline-controlled transactivator (TTA) is expressed specifically in insulin-producing cells to ensure regulated TGFβ expression in TTA/TGFβ NOD mice, as previously described (Peng et al. 2004). NOD transgenic mice were fed with normal or doxycycline-supplemented food and monitored for disease development.

9.3   Results

9.3.1   TGFβ and Effector T Cells

Mice with T cell-specific blockade of TGFβ signaling developed immunopathology in multiple organs including lung, colon, and kidney. Both $CD4^+$ and $CD8^+$ T cells from DNR mice differentiated readily into effector/memory T cells and secreted cytokines upon in vitro stimulation. Levels of T-dependent classes of immunoglobulins in the sera were consistently found to be increased. These observations revealed an essential function for TGFβ in T cell tolerance (Gorelik and Flavell 2000). To investigate how TGFβ regulates tumor immunity, we challenged mice with syngeneic B16-F10 or EL4 tumor cells. In both cases, DNR mice were found to be resistant to the tumor challenge. Blockade of TGFβ signaling in $CD8^+$ T cells was essential for this protection, and was associated with the expansion of tumor-specific $CD8^+$ T cells (Gorelik and Flavell 2001). $CD4^+$ T cells play a critical role in orchestrating both the humoral and cellular arms of immune responses. $CD4^+$ T cells from BALB/c mice selectively differentiate to a skewed Th2 phenotype, which renders these mice susceptible to infection by intracellular pathogens such as *Leishmania*. To study the function of TGFβ signaling in pathogen infection, DNR mice were backcrossed to the

BALB/c background and investigated for their response to *Leishmania major* infection. Significantly, DNR mice were found to be resistant to the infection, because they developed an enhanced Th1-type cytokine response. Interestingly, Th2 cytokines were also found to be elevated in the DNR mice (Gorelik et al. 2000). These observations suggested that TGFβ inhibits both arms of effector T cell differentiation. T helper cell differentiation is regulated by environmental cues and associated with epigenetic changes in the cytokine loci. Master regulators including GATA-3 for Th2 and T-bet for Th1 differentiation have been identified by us (Zheng and Flavell 1997) and Szabo et al. (2000). Significantly, TGFβ inhibited the expression of both factors which explains its regulation of help T cell differentiation (Gorelik et al. 2000, 2002).

9.3.2 TGFβ and Regulatory T Cells

Several studies have demonstrated that $CD4^+CD25^+$ regulatory T cells produce elevated levels of TGFβ (Green et al. 2003). The fact that enhanced TGFβ signaling receptors reside on the membrane of $CD4^+CD25^+$ regulatory T cells underscores the potential for autocrine and/or paracrine receptor–ligand interactions in these cells. In this study, we provide direct evidence that TGFβ is a positive regulator of $CD4^+CD25^+$ regulatory T cell expansion in vivo. We generated mice in which TGFβ expression can be induced temporally by the tetracycline regulatory system (Peng et al. 2004). Data from RT-PCR and histochemistry studies showed that TGFβ gene and protein expression can be efficiently turned on and off in the islets within 1 week after changing the diet to a doxycycline-containing food source. Using this system, we could control and target the expression of the transgene at specific stages of diabetes development. Transgenic mice from all groups were monitored for diabetes development in comparison to transgene-negative control littermates for 60 weeks. We found that TGFβ expression completely blocked the development of diabetes in NOD transgenic mice. Furthermore, a short pulse of TGFβ in the islets during the priming phase of the disease was sufficient to provide protection by promoting the expansion of the intra-islet $CD4^+CD25^+$ T cell pool. Approximately 40%–50% of intra-islet $CD4^+$ T cells expressed the CD25 marker and exhibited characteristics of regulatory T cells, including small size, high level of intracellular CTLA-4,expression of Foxp3, and the ability to transfer protection against diabetes. Results from in vivo incorporation of BrdUrd revealed that the generation of a high frequency of regulatory T cells in the islets is due to in situ expansion as a result of TGFβ expression. These findings demonstrated that TGFβ inhibits autoimmune diseases via regulation of the size of the CD4$^+$CD25$^+$ regulatory T cell pool in vivo, a previously uncharacterized mechanism.

9.4 Conclusions

Understanding the role of TGFβ signaling in T-cell regulation has come a long way since the initial discovery of its ability to regulate T cell proliferation. Our recent work has demonstrated that TGFβ downregulates effector T cell differentiation by inhibiting the expression of transcription factors T-bet and GATA-3. The role of TGFβ in regulatory T cell biology warrants further investigation. Insights into the biology of regulatory T cells might provide important insights into how the different roles of TGFβ in regulatory T-cell function relate to heterogeneity of the regulatory T cell population or another regulatory T cell property. The complex role of TGFβ in the regulation of immune homeostasis is likely to fascinate investigators as they decipher the many mysteries of this family of molecules for many years to come (Fig. 1).

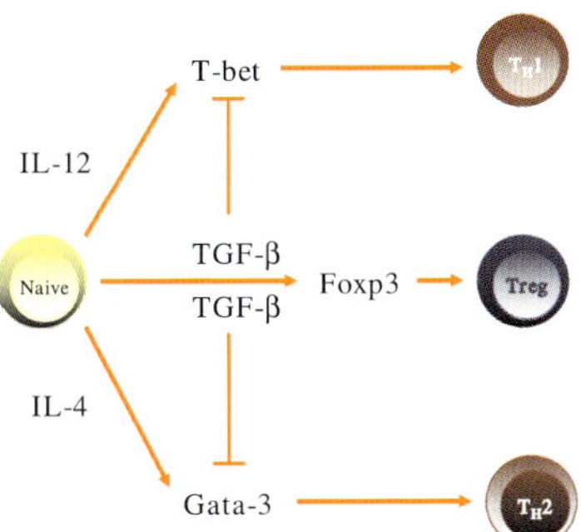

Fig. 1. Regulation of effector T cell and Treg cell differentiation by TGFβ. TGFβ prevents Th1 and Th2 cell differentiation through inhibition of T-bet and GATA-3 expression. TGFβ induces the expression of Foxp3 gene and promotes Treg cell differentiation

                                                Y. Peng et al.

References

Gorelik L, Flavell RA (2000) Abrogation of TGFbeta signaling in T cells leads to spontaneous T cell differentiation and autoimmune disease. Immunity 12:171–181

Gorelik L, Flavell RA (2001) Immune-mediated eradication of tumors through the blockade of transforming growth factor-beta signaling in T cells. Nat Med 7:1118–1122

Gorelik L, Fields PE, Flavell RA (2000) Cutting edge: TGF-beta inhibits Th type 2 development through inhibition of GATA-3 expression. J Immunol 165:4773–4777

Gorelik L, Constant S, Flavell RA (2002) Mechanism of transforming growth factor beta-induced inhibition of T helper type 1 differentiation. J Exp Med 195:499–1505

Green EA, Gorelik L, McGregor CM et al (2003) CD4$^+$CD25$^+$ T regulatory cells control anti-islet CD8$^+$ T cells through TGF-beta-TGF-beta receptor interactions in type 1 diabetes. Proc Natl Acad Sci U S A 100:10878–10883

Peng Y, Laouar Y, Li MO et al, (2004) TGF-beta regulates in vivo expansion of Foxp3-expressing CD4$^+$CD25$^+$ regulatory T cells responsible for protection against diabetes. Proc Natl Acad Sci U S A 101:4572–4577

Shull MM, Ormsby I, Kier AB et al (1992) Targeted disruption of the mouse transforming growth factor-beta 1 gene results in multifocal inflammatory disease. Nature 359(6397):693–699

Szabo SJ, Kim ST, Costa GL, Zhang X, Fathman CG, Glimcher LH (2000) A novel transcription factor, T-bet, directs Th1 lineage commitment. Cell. 100:655–669

Zheng W, Flavell RA (1997) The transcription factor GATA-3 is necessary and sufficient for Th2 cytokine gene expression in CD4 T cells. Cell. 89:587–96

10 TNF Blockade: An Inflammatory Issue

B.B. Aggarwal, S. Shishodia, Y. Takada, D. Jackson-Bernitsas,
K.S. Ahn, G. Sethi, H. Ichikawa

10.1 Introduction 162
10.2 TNF Cell Signaling 162
10.3 Inhibitors of TNF Cell Signaling 166
10.3.1 NF-κB Blockers 166
10.3.2 AP-1 Blockers 168
10.3.3 Suppression of TNF-Induced P38 MAPK Activation 168
10.3.4 Suppression of TNF-Induced JNK Activation 169
10.3.5 Suppression of TNF-Induced P42/p44 MAPK Activation 170
10.3.6 Suppression of TNF-Induced AKT Activation 171
10.4 Role of TNF in Skin Diseases 172
10.5 Bright Side of TNF 174
10.6 Dark Side of TNF Blockers 174
10.7 Identification of Novel Blockers of TNF 178
10.8 Conclusions 180
References 180

Abstract. Tumor necrosis factor (TNF), initially discovered as a result of its antitumor activity, has now been shown to mediate tumor initiation, promotion, and metastasis. In addition, dysregulation of TNF has been implicated in a wide variety of inflammatory diseases including rheumatoid arthritis, Crohn's disease, multiple sclerosis, psoriasis, scleroderma, atopic dermatitis, systemic lupus erythematosus, type II diabetes, atherosclerosis, myocardial infarction, osteoporosis, and autoimmune deficiency disease. TNF, however, is a critical component of effective immune surveillance and is required for proper proliferation and function of NK cells, T cells, B cells, macrophages, and dendritic cells. TNF activity can be blocked, either by using antibodies (Remicade and Humira) or soluble TNF receptor (Enbrel), for the symptoms of arthritis and Crohn's disease to be alleviated, but at the same time, such treatment increases the risk of infections, certain type of cancers, and cardiotoxicity. Thus blockers of TNF that are safe and yet efficacious are urgently needed. Some evidence suggests that while the transmembrane form of TNF has beneficial effects, soluble TNF mediates toxicity. In most cells, TNF mediates its effects through activation of caspases, NF-κB, AP-1, c-jun N-terminal kinase, p38 MAPK, and p44/p42 MAPK. Agents that can differentially regulate TNF expression or TNF signaling can be pharmacologically safe and effective therapeutics. Our laboratory has identified numerous such agents from natural sources. These are discussed further in detail.

10.1 Introduction

Tumor necrosis factor (TNF)-α and TNF-β, produced primarily by monocytes and lymphocytes, respectively, were first isolated in 1984, as cytokines that kill tumor cells in culture and induce tumor regression in vivo (Aggarwal et al. 1984). Intravenous administration of TNF to cancer patients produced numerous toxic reactions, including fever (Kurzrock et al. 1985). In animal studies, TNF has been shown to mediate endotoxin-mediated septic shock (Beutler et al. 1985). Other reports have indicated that dysregulation of TNF synthesis mediates a wide variety of diseases, including rheumatoid arthritis and inflammatory bowel disease (also called Crohn's disease) (Fig. 1).

10.2 TNF Cell Signaling

TNF is a transmembrane protein with a molecular mass of 26 kDa that was originally found to be expressed in macrophages and has now been found to be expressed by a wide variety of cells. In response to various stimuli, TNF is secreted by the cells as a 17-kDa protein through a highly regulated process that involves an enzyme: TNF-activating converting enzyme (TACE).

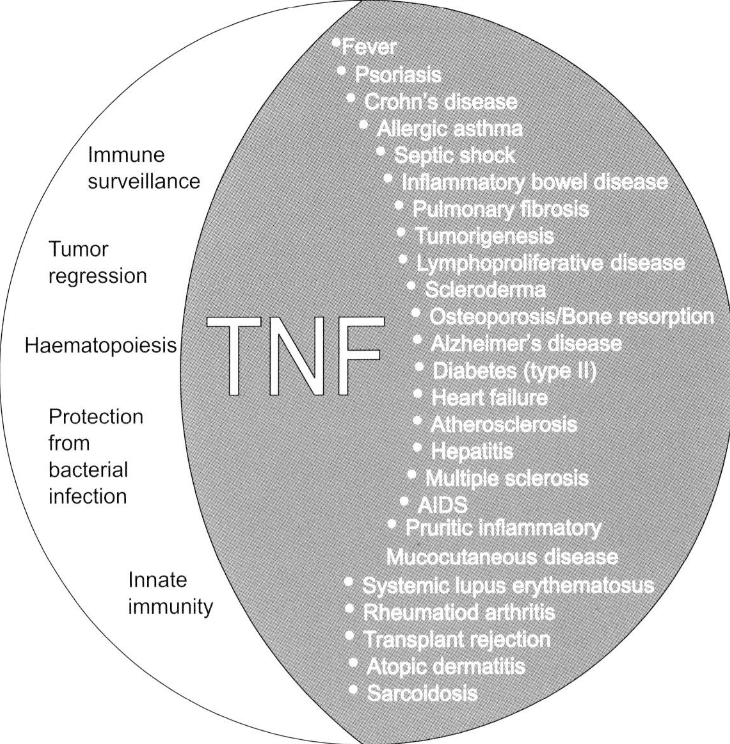

Fig. 1. TNF synthesis mediates a wide variety of diseases including rheumatoid arthritis and inflammatory bowel disease

TNF mediates its effects through two different receptors (Fig. 2): TNF receptor I (also called p55 or p60) and TNF receptor II (also called p75 or p80). While TNF receptor I is expressed on all cell types in the body, TNF receptor II is expressed selectively on endothelial cells and on cells of the immune system. TNF binds to two receptors with comparable affinity. Why there are two different receptors for TNF is incompletely understood. Evidence related to differential signaling (Aggarwal et al. 2002), ligand passing (Tartaglia et al. 1993), binding to soluble TNF vs transmembrane TNF (Grell et al. 1995) has been presented.

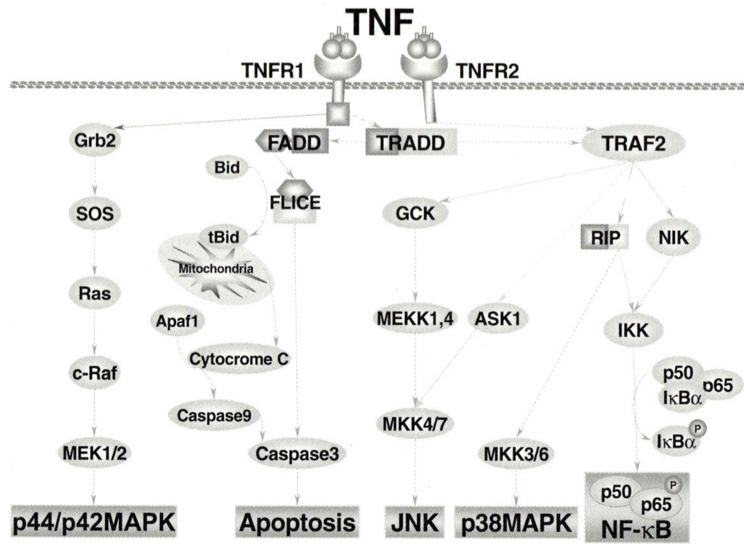

Fig. 2. TNF cell signaling pathway

The cytoplasmic domain of the TNF receptor I has a death domain, which has been shown to sequentially recruit TNF receptor-associated death domain (TRADD), Fas-associated death domain (FADD), and FADD-like interleukin-1β-converting enzyme (FLICE) (also called caspase-8), leading to caspase-3 activation, which in turn induces apoptosis by inducing degradation of multiple proteins (Nagata and Golstein 1995). TRADD also recruits TNF receptor-associated factor (TRAF)2, which through receptor-interacting protein (RIP) activates IκBα kinase (IKK), leading to IκBα phosphorylation, ubiquitination, and degradation, which finally leads to nuclear factor-κB (NF-κB) activation. NF-κB activation is followed by the expression of various genes that can suppress the apoptosis induced by TNF. Through recruitment of TRAF2, TNF has also been shown to activate various mitogen-activated protein kinases (MAPK), including the c-jun N-terminal kinases (JNK) p38 MAPK and p42/p44 MAPK. TRAF2 is also essential for the TNF-induced activation of AKT, another cell-survival signaling pathway. Thus TNFRI activates both apoptosis and cell survival signaling pathways simultaneously.

In contrast to TNFR1, the cytoplasmic domain of TNFR2 lacks the death domain and binds TRAF1 and TRAF2 directly. Through activation of JNK, TNF activates AP-1, another redox-sensitive transcription factor. Gene-deletion studies have shown that TNFR2 can also activate NF-κB, JNK, p38 MAPK, and p42/p44 MAPK (Mukhopadhyay et al. 2001).

TNFR2 can also mediate TNF-induced apoptosis (Haridas et al. 1998). Because TNFR2 cannot recruit TRADD-FADD-FLICE, how TNFR2 mediates apoptosis is not understood. Various pieces of evidence suggest that homotrimeric TNF binds to homotrimeric TNF receptor to mediate its signals (Ameloot et al. 2001). TNF receptor deletion studies have provided evidence that this receptor communicates with receptors for other ligands, including receptor activator of NF-κB ligand (RANKL, a member of the TNF superfamily) and endotoxin (Takada and Aggarwal 2003b, 2004).

Since its discovery, TNF has been linked to a wide variety of diseases. How TNF mediates disease-causing effects is incompletely understood. The induction of pro-inflammatory genes by TNF has been linked to most diseases. The pro-inflammatory effects of TNF are primarily due to its ability to activate NF-κB. Almost all cell types, when exposed to TNF, activate NF-κB, leading to the expression of inflammatory genes. Over 200 genes have been identified that are regulated by NF-κB activation (Kumar et al. 2004). These include cyclooxygenase-2 (COX-2), lipoxygenase-2 (LOX-2), cell-adhesion molecules, inflammatory cytokines, chemokines, and inducible nitric oxide synthase (iNOS). TNF mediates some of its disease-causing effects by modulating growth. For instance, for most tumor cells TNF has been found to be a growth factor (Sugarman et al. 1985). These include ovarian cancer cells, cutaneous T cell lymphoma (Giri and Aggarwal 1998) glioblastoma (Aggarwal et al. 1996), acute myelogenous leukemia (Tucker et al. 2004), B cell lymphoma (Estrov et al. 1993), breast carcinoma (Sugarman et al. 1987), renal cell carcinoma (Chapekar et al. 1989), multiple myeloma (Borset et al. 1994), and Hodgkin's lymphoma (Hsu and Hsu 1990). Various fibroblasts, including normal human fibroblasts, scleroderma fibroblasts, synovial fibroblasts, and periodontal fibroblasts, proliferate in response to TNF. Why on treatment with TNF some cells undergo apoptosis, others undergo proliferation, and most are unaffected is not understood. The differences are not due to lack of receptors or variations in their affinity.

10.3   Inhibitors of TNF Cell Signaling

Because of the critical role of TNF in mediating a wide variety of diseases, TNF has become an important target for drug development. Since TNF mediates its effects through activation of NF-κB, AP-1, JNK, p38 MAPK, p44/p42 MAPK, and AKT (Fig. 3); agents that can suppress these pathways have potential for therapy of TNF-linked diseases. Below is a review of inhibitors of such pathways.

10.3.1   NF-κB Blockers

Most diseases that have been linked to TNF have also been linked to NF-κB activation (Aggarwal 2003), indicating that TNF mediates its pathological effects through activation of NF-κB. Thus blockers of NF-κB have a potential for alleviating TNF-linked diseases. Several NF-κB blockers have been identified using targets that mediate the TNF-induced NF-κB activation pathway (Fig. 4). Various hormones and cytokines have also been described that can suppress TNF-induced NF-κB activation. These include IL-4, IL-13, IL-10, melanocyte-stimulating hormone (βMSH), luteinizing hormone (LH), human chorionic gonadotrophin (HCG), and IFN-α/β (Aggarwal et al. 2002). Other agents that suppress NF-κB activation include inhibitors of proteosomes, inhibitors of

Signaling pathways activated by TNF

Inflammatory pathway	NF-κB
Stress pathway	JNK & p38
Cell survival pathway	PI-3K/Akt
Mitogenic pathway	MAPK/Erk
JAK/STAT pathway	
Apoptosis pathway	Caspases

Fig. 3. TNF mediates its effects through activation of NF-κB, AP-1, JNK, p38 MAPK, p44/p42 MAPK, and AKT

Potential strategies to suppress TNF production and signaling through inhibition of NF-κB activation

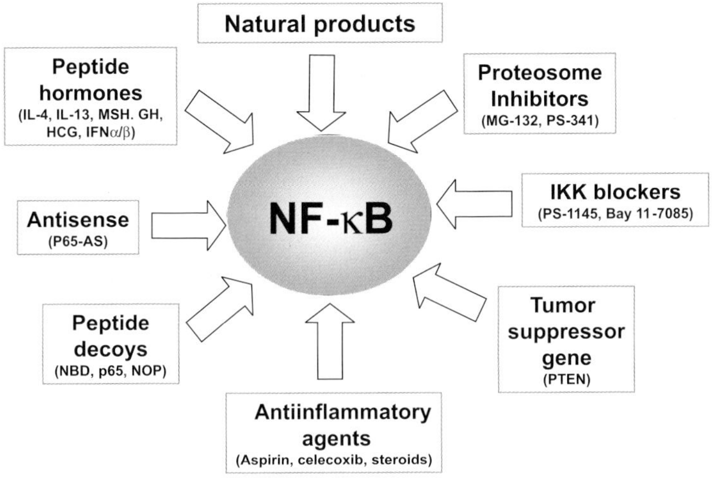

Fig. 4. Potential strategies to suppress TNF production and signaling through inhibition of NF-κB activation

ubiquitination, inhibitors of the protein kinase that phosphorylates IκBα (IKK), inhibitors of IKK activation (Fig. 5), decoy peptides of IKK and p65, antisense oligonucleotides, and Si RNA to p65. These inhibitors, although quite effective in suppressing NF-κB activation, have shown limited promise in the treatment of the disease for a variety of reasons. These include in vivo toxicity, lack of specificity, and poor bioavailability. Thus safer and efficacious blockers of TNF-induced NF-κB activation are needed. Our laboratory has described numerous agents derived from natural sources that can suppress NF-κB activation induced by TNF quite effectively. Through suppression of NF-κB, these inhibitors can also suppress expression of TNF. These polyphenols (e.g., curcumin, a diferuloylmethane) have been tested in animals and found to exhibit good activity in suppression of tumorigenesis, Alzheimer disease, diabetes, arthritis, Crohn's disease, and cardiovascular disease.

Development of IKK blockers

Compound	IC50 (in vitro)	IC50 (in vivo)	Investigator
Bayer Yakuhin BY-4j	8.5 nM	0.06uM	Murata T, 2004
Millennium PS1145/MNL120	20 nM	3 uM	Hideshima T, 2002
Celgene SPC839	62 nM	10 uM	Palanki MS, 2003
BMS BMS345541	300 nM	1-5 uM	Burke et al, 2003
Pharmacia/Pfizer SC-514	3-12 uM	100 uM	Kishore N, 2003
Wyeth-Ayerst WAY-169916	1 uM	N. A.	Steffan R, 2004

SPC839 is now called AS602868

Fig. 5. Development of IKK blockers

10.3.2   AP-1 Blockers

Although TNF is a potent activator of AP-1, no specific blocker of AP-1 activation has yet been described. However, most phytochemicals that suppress NF-κB activation also suppress AP-1 activation (Manna and Aggarwal 2000; Manna et al. 2000; Torii et al. 2004).

10.3.3   Suppression of TNF-Induced P38 MAPK Activation

TNF is one of the most potent activators of p38 MAPK (Lee et al. 1994). This activation is specifically inhibited by SB203580 (Henry et al. 1998; Kumar et al. 1997). Campbell et al. (2004) examined the role of p38 MAPK in the regulation of TNF in primary human cells relevant to inflammation, e.g., macrophages and rheumatoid synovial cells (Campbell et al. 2004). Using a dominant-negative variant (D168A) of p38 MAPK and a kinase inhibitor, SB203580, they confirmed in primary human macrophages that p38 MAPK regulates TNF production using a posttranscriptional mechanism requiring the $3'$ untranslated region of the gene. However, in LPS-activated primary human macrophages, they found p38 MAPK modulation of TNF transcription was mediated through p38 MAPK regulation of NF-κB. Interestingly, this mechanism was not observed in rheumatoid synovial cells. It is important to note that the dominant negative mutant of p38 MAPK, but not SB203580, was effective at inhibiting spontaneous TNF production in these ex vivo rheumatoid synovial cell cultures. These data indicate that there are potential major differences in the role of p38 MAPK in inflammatory signaling that have a bearing on the use of this kinase as a target for therapy. These results also indicate that this kinase is a valid target in rheumatoid disease. Vanden Berghe et al. (1998) found that p38 and p44/p42 MAPK pathways are required for NF-κB p65 transactivation mediated by TNF. De Alvaro et al. (2004) showed that TNF produces insulin resistance in skeletal muscle by activation of IKK in a p38 MAPK-dependent manner. Similarly, Hernandez et al. (2004) showed that Rosiglitazone ameliorates insulin resistance in brown adipocytes of Wistar rats by impairing TNF induction of p38 and p42/p44 MAPK.

10.3.4 Suppression of TNF-Induced JNK Activation

TNF is one of the most potent activators of JNK (Chen and Goeddel 2002). This activation is specifically inhibited by SP600125. Bennett et al. (2001) reported the identification of an anthrapyrazolone series that significantly inhibits JNK1, -2, and -3 (K(i) = 0.19 μM). SP600125 is a reversible ATP-competitive inhibitor with greater than 20-fold selectivity vs a range of kinases and enzymes tested. In cells, SP600125 dose dependently inhibited the phosphorylation of c-Jun and the expression of the inflammatory genes COX-2, IL-2, IFN-γ, and TNF and prevented the activation and differentiation of primary human CD4 cell cultures. In animal studies, SP600125 blocked LPS-induced expression of TNF and inhibited anti-CD3-induced apoptosis of CD4(+) CD8(+) thymocytes. Our study supports targeting JNK as an important strategy in inflammatory disease, apoptotic cell death, and cancer. Ishii et al. (2004) found that inhibition of JNK activity improves ischemia/reperfusion injury in rat lungs.

10.3.5 Suppression of TNF-Induced P42/p44 MAPK Activation

Besides p38 MAPK and JNK, TNF is also potent activator of p42/p44 MAPK, also called extracellular signal-regulated kinase (ERK)1/2 (Johnson and Lapadat 2002). This activation is inhibited by PD98059 (Kumar et al. 1998b). TNF-induced ERK activation is required for proliferation of most cells. Murakami-Mori et al. (1999) found that ERK1/2 in Kaposi sarcoma (KS) cells was significantly activated by TNF through tyrosine/threonine phosphorylation. A selective inhibitor for ERK1/2 activator kinases, PD98059, profoundly inhibited not only the activation of ERK1/2, but also TNF-induced KS cell proliferation. They therefore proposed that the TNFRI-ERK1/2 pathway plays a pivotal role in transmitting to KS cells the mitogenic signals of TNF. They found that actinomycin D treatment of KS cells selectively abolished expression of MADD, a novel TNFRI-associated death domain protein. TNF-α failed to induce ERK1/2 activation in the actinomycin D-treated cells. MADD may couple TNFRI with the ERK1/2 signaling pathway required for KS cell proliferation. Using similar inhibitors, Goetze et al. (1999) found that TNF-induced migration of vascular smooth muscle cells is MAPK-dependent. Tran et al. (2001) found that MAPK/ERK overrides the apoptotic signaling from Fas, TNF, and TRAIL receptors. They observed that a number of FasR-insensitive cell lines could redirect the proapoptotic signal to an anti-apoptotic ERK1/2 signal, resulting in inhibition of caspase activation. They determined that similar mechanisms are operational in regulating the apoptotic signaling of other death receptors. Activation of the FasR, TNFR1, and TRAILR rapidly induced ERK1/2 activation, an event independent of caspase activity. Whereas inhibition of the death receptor-mediated ERK1/2 activation was sufficient to sensitize the cells to apoptotic signaling from FasR and TRAIL-R, cells were still protected from apoptotic TNFR1 signaling. The latter seemed to be attributable to the strong activation of the anti-apoptotic factor NF-κB, which remained inactive in FasR or TNF-related apoptosis-inducing ligand receptor (TRAILR) signaling. However, when the cells were sensitized with cycloheximide, which is sufficient to sensitize the cells also to apoptosis by TNFR1 stimulation, Tran et al. noticed that adenovirus-mediated expression of constitutively active MKK1 could rescue the cells from apoptosis induced by the receptors by preventing caspase-8 activation. Taken together, these results showed that ERK1/2 has a dominant protecting effect over apoptotic signaling from the death receptors. This protection, which is independent of newly synthesized proteins, acts in all cases by suppressing activation of the caspase effector machinery.

10.3.6 Suppression of TNF-Induced AKT Activation

TNF is also a potent activator of AKT. Ozes found that NF-κB activation by TNF requires the Akt serine-threonine kinase (Ozes et al. 1999). Pastorino found TNF induced the phosphorylation of BAD by Akt at serine 136 in HeLa cells under conditions that are not cytotoxic. BAD phosphorylation by TNF was dependent on phosphatidylinositide-3-OH kinase (PI3K) and was accompanied by the translocation of BAD from the mitochondria to the cytosol (Pastorino et al. 1999). Blocking the phosphorylation of BAD and its translocation to the cytosol with the PI3K inhibitor wortmannin activated caspase-3 and markedly potentiated the cytotoxicity of TNF. Transient transfection with a PI3K dominant-negative mutant or a dominant-negative mutant of the serine-threonine kinase Akt, the downstream target of PI3K, and the enzyme that phosphorylates BAD similarly potentiated the cytotoxicity of TNF. By contrast, transfection with a constitutively active Akt mutant protected against the cytotoxicity of TNF in the presence of wortmannin. Phosphorylation of BAD prevents its interaction with the antiapoptotic protein Bcl-X_L. Transfection with a Bcl-X_L expression vector protected against the cytotoxicity of TNF in the presence of wortmannin.

Yang et al. (2004) reported the discovery of a small-molecule Akt pathway inhibitor, Akt/protein kinase B signaling inhibitor-2 (API-2), by screening the National Cancer Institute Diversity Set. API-2 suppressed the kinase activity and phosphorylation level of Akt. The inhibition of Akt kinase resulted in suppression of cell growth and induction of apoptosis in human cancer cells. API-2 was highly selective for Akt and did not inhibit the activation of PI3K, phosphoinositide-dependent kinase-1, protein kinase C, serum- and glucocorticoid-inducible kinase, protein kinase A, signal transducer and activators of transcription 3, ERK1/2, or JNK. Furthermore, API-2 potently inhibited tumor growth in nude mice of human cancer cells in which Akt is aberrantly expressed/activated but not of those cancer cells in which it is not.

10.4   Role of TNF in Skin Diseases

Although it suppresses the proliferation of cells, TNF was also found to induce the proliferation of dermal fibroblasts (Berman and Wietzerbin 1992). The TNF-induced JNK pathway has been found to be dysregulated in patients with familial cylindromatosis in whom CYLD, a tumor suppressor, is mutated. Such patients have an autosomal dominant predisposition to multiple tumors of the skin of the arms and legs (Reiley et al. 2004). Dysregulation of TNF has been linked with several skin diseases, including rheumatoid arthritis, atopic dermatitis, scleroderma, psoriasis (Kane and FitzGerald 2004), systemic sclerosis, Crohn's disease, pyoderma gangrenosum, sarcoidosis, and cutaneous T cell lymphoma.

Furthermore, keratinocyte-derived TNF acts as an endogenous tumor promoter and can also regulate AP-1 activity in mouse epidermis. Arnott found that TNFR2 cooperated with TNFR1 to optimize TNFR1-mediated TNF bioactivity on keratinocytes in vitro. They found that expression of both TNF-receptor subtypes is essential for optimal skin tumor development and provided some rationale for the use of TNF antagonists in the treatment of cancer (Arnott et al. 2004).

Because keloids, which are characterized as an overexuberant healing response, represent an inflammatory response, it is logical to assume that cytokines play a part in orchestrating keloid pathology. Messadi identified differences in the expression of NF-κB and its related genes between keloid and normal skin fibroblasts (Messadi et al. 2004). They showed that TNF upregulated 15% of NF-κB signal pathway-related genes in keloid fibroblasts compared to normal skin. At the protein level, keloid fibroblasts and tissues showed higher basal levels of the TNF-receptor-associated factors TRAF1, TRAF2, inhibitor of apoptosis (c-IAP-1), and NF-κB, compared with normal skin fibroblasts. Keloid fibroblasts showed a constitutive increase in NF-κB-binding activity both with and without TNF-α treatment. It is possible that NF-κB and its targeted genes, especially the antiapoptotic genes, could play a role in keloid pathogenesis; thus targeting NF-κB could help in developing therapeutic interventions for the treatment of keloid scarring.

TNF receptor-associated periodic syndrome (TRAPS) is an autosomal dominant inherited condition characterized by periodic fever and pain. TRAPS is a model for a novel pathogenic concept in which a TNF receptor fails to be shed; thus suggesting that medications targeting TNF may be effective in TRAPS (Masson et al. 2004).

That TNF signaling plays a role in skin is also evident from molecular imaging of NF-κB, a primary regulator of stress response. Carlsen developed transgenic mice that express luciferase under the control of NF-κB, enabling real-time noninvasive imaging of NF-κB activity in intact animals (Carlsen et al. 2004). We showed that, in the absence of stimulation, strong, intrinsic luminescence is evident in lymph nodes in the neck region, thymus, and Peyer's patches. Treating mice with stressors such as TNF-α, IL-1α, or LPS increases the luminescence in a tissue-specific manner, with the strongest activity observable in the skin, lungs, spleen, Peyer's patches, and the wall of the small intestine. Liver, kidney, heart, muscle, and adipose tissue exhibit less intense activities. Exposure of the skin to a low dose of UV-B radiation increases luminescence in the exposed areas. In ocular experiments, when LPS and TNF-α were injected into NF-κB-luciferase transgenic mice, a 20- to 40-fold increase in NF-κB activity occurred in the lens, as well as in other LPS- and TNF-α-responsive organs. Peak NF-κB activity occurred 6 h after injection of TNF-α and 12 h after injection of LPS.

Mice exposed to 360 J/m(2) of UV-B exhibited a 16-fold increase in NF-κB activity 6 h after exposure, much like TNF-α-exposed mice. Thus, in NF-κB-luciferase transgenic mice, NF-κB activity also occurs in lens epithelial tissue and is activated when the intact mouse is exposed to classical stressors. Furthermore, as revealed by real-time noninvasive imaging, induction of chronic inflammation resembling rheumatoid arthritis produces strong NF-κB activity in the affected joints. These investigators used this model to demonstrate regulation by manipulating vitamin A status in mice. NF-κB activity is elevated in mice fed a vitamin A-deficient (VAD) diet and suppressed by excess doses of retinoic acid. They thus demonstrated the development and use of a versatile model for monitoring NF-κB activation both in tissue homogenates and in intact animals after the use of classical activators, during disease progression and after dietary intervention.

10.5 Bright Side of TNF

Two TNF-neutralizing agents are licensed in the US. Infliximab has been licensed for the treatment of Crohn's disease and is used with methotrexate for the treatment of rheumatoid arthritis (Hanauer 2004). Etanercept is a soluble TNF receptor type 2 that is licensed for the treatment of rheumatoid arthritis, including the juvenile form, and, more recently, was licensed for the treatment of psoriatic arthritis (Baraliakos and Braun 2004). Anti-TNF does not cure rheumatoid arthritis or Crohn's disease, but blocking TNF may reduce the inflammation caused by too much TNF.

10.6 Dark Side of TNF Blockers

Serious infections, including sepsis and fatal infections, have been reported in patients receiving TNF-blocking agents. Many of the serious infections in patients treated with anti-TNF have occurred in patients on concomitant immunosuppressive therapy that, in addition to their Crohn's disease or rheumatoid arthritis, could predispose them to infections. Caution should be exercised when considering the use of anti-TNF in patients with a chronic infection or a history of recurrent infection. Anti-TNF should not be given to patients with a clinically important, active infection. The companies that manufacture TNF blockers have indicated serious side effects of these drugs (Table 1).

Some patients who took Remicade developed symptoms that can resemble a disease called lupus. Lupus-like symptoms may include chest discomfort or pain that does not go away, shortness of breath, joint pain, or a rash on the cheeks or arms that gets worse in the sun.

Although TNF blocking therapy is currently used, a great deal of caution is needed. Mohan's group described the clinical features of leukocytoclastic vasculitis (LCV) associated with the use of TNF-alpha blockers (Mohan et al. 2004). They identified 35 cases of LCV, 20 following etanercept administration and 15 following infliximab administration. Seventeen of the 35 (48.5%) were biopsy-proven cases, and the others had skin lesions that were clinically typical for LCV. Twenty-two of 35 (62.8%) patients had complete or marked improvement of skin lesions upon stopping the TNF-α blocker. Three patients who had re-

Table 1. Side effects associated with administration of TNF blockers in human

Infections	
Granulomatous infection	Wallis et al. 2003
Opportunistic infection	Slifman et al. 2003
	Lee et al. 2002
Heart failure	
Hematopoiesis	
Allergic reactions	
Nervous system disorders	
Lymphoma	Wolfe and Michaud 2004
	Sandborn et al. 2004
Lupus-like symptoms	
Leukocytoclastic vasculitis (LCV)	Mohan et al. 2004
	Devos et al. 2003

ceived etanercept had continuing lesions despite discontinuation of the drug; one of these patients improved when switched to infliximab. One patient who received infliximab was reported to have continuing lesions despite discontinuation of the drug and treatment with prednisone and antihistamines. Six patients experienced a positive rechallenge (recurrence of LCV on restarting therapy with a TNF-α blocker) and three patients a negative rechallenge phenomenon. LCV lesions improved in patients despite continuing use of concomitant medications reportedly associated with LCV.

Various adverse cutaneous reactions to anti-TNF-α monoclonal antibody have been reported. In clinical studies with infliximab (Remicade), adverse drug reactions were most frequently reported in the respiratory system and in the skin and appendages. Devos described six patients receiving anti-TNF-α therapy (infliximab) for Crohn's disease or rheumatoid arthritis who showed adverse cutaneous reactions and were diagnosed with LCV, lichenoid drug reaction, perniosis-like eruption (two patients), superficial granuloma annulare, and acute folliculitis (Devos et al. 2003).

Wallis's group examined the relationship between the use of TNF antagonists and onset of granulomatous infection using data collected through the Adverse Event Reporting System of the US Food and Drug Administration for January 1998–September 2002 (Wallis et al. 2003). Granulomatous infections were reported at rates of approximately 239 per 100,000 patients who received infliximab and approximately 74 per 100,000 patients who received etanercept, indicating a significant difference between the two drugs. Tuberculosis was the most frequently reported disease, occurring in approximately 144 and approximately 35 per 100,000 infliximab-treated and etanercept-treated patients, respectively. Candidiasis, coccidioidomycosis, histoplasmosis, listeriosis, and other infections caused by nontuberculous mycobacteria were reported with significantly greater frequency among infliximab-treated patients. Seventy-two percent of these infections occurred within 90 days after starting infliximab treatment, and 28% occurred after starting etanercept treatment ($p < 0.001$). These data indicate a risk of granulomatous infection that was 3.25-fold greater among patients who received infliximab than among those who received etanercept. The clustering of reports shortly after initiation of treatment with infliximab is consistent with reactivation of latent infection.

Because of the potential for a decrease in host resistance to infectious agents due to treatment with anti-TNF agents, Slifman evaluated cases of opportunistic infection, including those caused by *Listeria monocytogenes*, in patients treated with these products (Slifman et al. 2003). The FDA Adverse Event Reporting System, a passive monitoring system, was reviewed to identify all reports of adverse events (through December 2001) associated with *L. monocytogenes* infection in patients treated with infliximab or etanercept. Fifteen cases associated with infliximab or etanercept treatment were identified. In 14 of these cases, patients had received infliximab. The median age of all patients was 69.5 years (range, 17–80 years); 53% were women. Six deaths were reported. Among patients for whom an indication for use was reported, nine patients (64%) with rheumatoid arthritis and five patients (36%) with Crohn's disease (information was not reported for one patient). All patients for whom information was reported were receiving concurrent immunosuppressant drugs. Thus postlicensure surveillance suggests that *L. monocytogenes* infection may be a serious complication of treatment with TNF-neutralizing agents, particularly infliximab.

Lee also sought to identify cases of opportunistic infection, including histoplasmosis, in patients treated with these products (Lee et al. 2002). The US Food and Drug Administration's passive surveillance database for monitoring postlicensure adverse events was reviewed to identify all reports received through July 2001 of histoplasmosis in patients treated with either infliximab or etanercept. Ten cases of *Histoplasma capsulatum* infection were reported: nine associated with infliximab and one associated with etanercept. In patients treated with infliximab, manifestations of histoplasmosis occurred within 1 week to 6 months after the first dose and typically included fever, malaise, cough, dyspnea, and interstitial pneumonitis. Of the ten patients with histoplasmosis, nine required treatment in an intensive care unit and one died. All patients had received concomitant immunosuppressive medications in addition to infliximab or etanercept, and all resided in *H. capsulatum*-endemic regions. Thus postlicensure surveillance suggests that acute life-threatening histoplasmosis may complicate immunotherapy with TNF-α antagonists, particularly infliximab. Histoplasmosis should be considered early in the evaluation of patients who reside in *H. capsulatum*-endemic areas in whom infectious complications develop during treatment with infliximab or etanercept.

The risk of lymphoma is increased in patients with rheumatoid arthritis, and some studies suggest that methotrexate and anti-TNF therapy might be associated independently with an increased risk of lymphoma. However, data from clinical trials and clinical practice do not provide sufficient evidence concerning these issues because of small sample sizes and selected study populations. Wolfe and Michaud (2004) determined the rate of and standardized incidence ratio (SIR) for lymphoma in patients with rheumatoid arthritis in general and in these patients by treatment group. Additionally, predictors of lymphoma in rheumatoid arthritis were determined. 18,572 patients with rheumatoid arthritis who were enrolled in the National Data Bank for Rheumatic Diseases (NDB) were prospectively studied. Patients were surveyed biannually, and potential lymphoma cases received detailed follow-up. The SEER (Survey, Epidemiology, and End Results) cancer data resource was used to derive the expected number of cases of lymphoma in a cohort that was comparable in age and sex. The overall SIR for lymphoma was 1.9 [95% confidence interval (95% CI), 1.3–2.7]. The SIR for biologic use was 2.9 (95% CI, 1.7–4.9) and for the use of infliximab (with or without etanercept) was 2.6 (95% CI, 1.4–4.5). For etanercept, with or without infliximab, the SIR was 3.8 (95% CI, 1.9–7.5). The SIR for MTX was 1.7 (95% CI, 0.9–3.2), and was 1.0 (95% CI, 0.4–2.5) for those not receiving MTX or biologics. Lymphoma was associated with increasing age, male sex, and education. Although the SIR is greatest for anti-TNF therapies, differences between therapies were slight, and confidence intervals for treatment groups overlapped. The increased lymphoma rates observed with anti-TNF therapy may reflect channeling bias, whereby patients with the highest risk of lymphoma preferentially receive anti-TNF therapy. Current data are insufficient to establish a causal relationship between the treatments and the development of lymphoma.

Sandborn and Loftus also noted that patients with moderate to severely active Crohn's disease treated with infliximab may have a small but real risk of developing severe infections, opportunistic infections, and non-Hodgkin's lymphoma (Sandborn et al. 2004).

10.7   Identification of Novel Blockers of TNF

Small molecules such as plant-derived phytochemicals, which are safer and yet effective in suppressing both production and action of TNF, should be explored. Inasmuch as TNF expression in most cells is regulated by the transcription factor NF-κB, agents that suppress this factor will also block TNF production. Phytochemicals such as curcumin, resveratrol, betulinic acid, ursolic acid, sanguanarine, capsaicin, gingerol, anethole, and eugenol, have been shown to suppress NF-κB (Aggarwal et al. 2004; Dorai and Aggarwal 2004; Han et al. 2001; Kumar et al. 1998a; Murakami et al. 2003; Shishodia et al. 2003; Takada and Aggarwal 2003a), and thus can suppress TNF production (Lukita-Atmadja et al. 2002) (Table 2). Such agents should be further explored.

Table 2. Phytochemicals known to suppress NF-κB activation induced by in-flammatory agents

Polyphenols	Terpenes
Amentoflavone	Andalusol
Apigenin	Anethol and analogs
Anethole	Artemisinin
Arctigenin and demethyltraxillagenin	Avicins
Baicalein and its derivatives	Betulinic acid
Bakuchiol (Drupanol)	Celastrol
Cannabinol	Costunolide
Capsaicinoids	Ergolide
Carnosol	Excisanin A
Catalposide	Foliol
Catechin and theaflavins	Germacranolides
Curcumin	Ginkgo biola ext.
Emodin	Ginsenoside Rg3
Flavopiridol	Guaianolides
Genistein	Helenalin
Glossogyne tenuifolia	Hypoestoxide
Hematein	Kamebacetal A
HMP	Kamebakaurin
Hypericin	Kaurenic acid
Isomallotochromanol and isomallotochromene	Linearol
Luteolin	Oleandrin
Nordihydroguaiaretic	Oxoacanthospermoldes
Acid	Parthenolide
Panduratin A	Pristimerin
Pycnogenol	Triptolide (PG 490)
Rhein	Ursolic acid
Rocaglamides	Alkaloids
Sanggenon C	Cepharanthine
Sphondin	Conophylline
Silymarin	Morphine and its analogs
Saucerneols	Tetrandine
Sauchinone and Manassantins	Sinomenine A
Wedelolactone	Benz[α]phenazine
Yakuchinones A and B	Lapachone
Benzopyrene	Caffeic acid phenethyl ester
Rotenone	CAPE
Chlorophyll catabolite	Phenolics
Pheophorbide A	Ethyl gallate
Iridoid glycoside	Saponin
Aucubin	Calagualine
Others	Stilbene
α-Lipoic acid	Resveratrol and analogues
Astaxanthin	
Germinated barley	
S-allylcysteine	
Vitamin C	
Vitamin E	

10.8   Conclusions

Extensive research in the last few years has clearly proven that TNF is a pro-inflammatory cytokine and thus is involved in the pathogenesis of variety of diseases. Suppression of TNF is a double-edged sword. While unquestionably TNF plays a critical role in inflammation, suppression of TNF will mitigate both its beneficial and its harmful effects. These studies show that safer modulators of TNF are needed.

Acknowledgements. Supported by the Clayton Foundation for Research (to BBA), Department of Defense US Army Breast Cancer Research Program grant (BC010610, to BBA), a PO1 grant (CA91844) from the National Institutes of Health on lung cancer chemoprevention (to BBA), a P50 Head and Neck Cancer SPORE grant from the National Institutes of Health (to BBA), and Cancer Center Core Grant CA 16672. We thank Mr. Walter Pagel for careful reading of the manuscript and providing valuable comments.

References

Aggarwal BB (2003) Signalling pathways of the TNF superfamily: a double-edged sword. Nat Rev Immunol 3:745–756

Aggarwal BB, Moffat B, Harkins RN (1984) Human lymphotoxin. Production by a lymphoblastoid cell line, purification, and initial characterization. Biol J Chem 259:686–691

Aggarwal BB, Schwarz L, Hogan ME, Rando RF (1996) Triple helix-forming oligodeoxyribonucleotides targeted to the human tumor necrosis factor (TNF) gene inhibit TNF production and block the TNF-dependent growth of human glioblastoma tumor cells. Cancer Res 56:5156–5164

Aggarwal BB, Shishodia S, Ashikawa K, Bharti AC (2002) The role of TNF and its family members in inflammation and cancer: lessons from gene deletion. Curr Drug Targets Inflamm Allergy 1:327–341

Aggarwal BB, Bhardwaj A, Aggarwal RS, Seeram NP, Shishodia S, Takada Y (2004) Role of resveratrol in prevention and therapy of cancer: preclinical and clinical studies. Anticancer Res 24:2783–2840

Ameloot P, Declercq W, Fiers W, Vandenabeele P, Brouckaert P (2001) Heterotrimers formed by tumor necrosis factors of different species or muteins. Biol J Chem 276:27098–27103

Arnott CH, Scott KA, Moore RJ, Robinson SC, Thompson RG, Balkwill FR (2004) Expression of both TNF-alpha receptor subtypes is essential for optimal skin tumour development. Oncogene 23:1902–1910

Baraliakos X, Braun J (2004) Current concepts in the therapy of the spondyloarthritides. BioDrugs 18:307–314

Bennett BL, Sasaki DT, Murray BW, O'Leary EC, Sakata ST, Xu W, Leisten JC, Motiwala A, Pierce S, Satoh Y, Bhagwat SS, Manning AM, Anderson DW (2001) SP600125, an anthrapyrazolone inhibitor of Jun N-terminal kinase. Proc Natl Acad Sci U S A 98:13681–13686

Berman B, Wietzerbin J (1992) Tumor necrosis factor-alpha (TNF-alpha), interferon-alpha (IFN-alpha) and interferon-gamma (IFN-gamma) receptors on human normal and scleroderma dermal fibroblasts in vitro. Dermatol J Sci 3:82–90

Beutler B, Milsark IW, Cerami AC (1985) Passive immunization against cachectin/tumor necrosis factor protects mice from lethal effect of endotoxin. Science 229:869–871

Borset M, Waage A, Brekke OL, Helseth E (1994) TNF and IL-6 are potent growth factors for OH-2, a novel human myeloma cell line. Eur Haematol J 53:31–37

Burke JR, Pattoli MA, Gregor KR, Brassil PJ, MacMaster JF, McIntyre KW, Yang X, Iotzova VS, Clarke W, Strnad J, Qiu Y, Zusi FC (2003) BMS-345541 is a highly selective inhibitor of I kappa B kinase that binds at an allosteric site of the enzyme and blocks NF-kappa B-dependent transcription in mice. J Biol Chem 278:1450–1456

Campbell J, Ciesielski CJ, Hunt AE, Horwood NJ, Beech JT, Hayes LA, Denys A, Feldmann M, Brennan FM, Foxwell BM (2004) A novel mechanism for TNF-alpha regulation by p38 MAPK: involvement of NF-kappaB with implications for therapy in rheumatoid arthritis. Immunol J 173:6928–6937

Carlsen H, Alexander G, Austenaa LM, Ebihara K, Blomhoff R (2004) Molecular imaging of the transcription factor NF-kappaB, a primary regulator of stress response. Mutat Res 551:199–211

Chapekar MS, Huggett AC, Thorgeirsson SS (1989) Growth modulatory effects of a liver-derived growth inhibitor, transforming growth factor beta 1, and recombinant tumor necrosis factor alpha, in normal and neoplastic cells. Exp Cell Res 185:247–257

Chen G, Goeddel DV (2002) TNF-R1 signaling: a beautiful pathway. Science 296:1634–1635

De Alvaro C, Teruel T, Hernandez R, Lorenzo M (2004) Tumor necrosis factor alpha produces insulin resistance in skeletal muscle by activation of inhibitor kappaB kinase in a p38 MAPK-dependent manner. Biol J Chem 279:17070–17078

Devos SA, Van Den Bossche N, De Vos M, Naeyaert JM (2003) Adverse skin reactions to anti-TNF-alpha monoclonal antibody therapy. Dermatology 206:388–390

Dorai T, Aggarwal BB (2004) Role of chemopreventive agents in cancer therapy. Cancer Lett 215:129–140

Estrov Z, Kurzrock R, Pocsik E, Pathak S, Kantarjian HM, Zipf TF, Harris D, Talpaz M, Aggarwal BB (1993) Lymphotoxin is an autocrine growth factor for Epstein-Barr virus-infected B cell lines. Exp J Med 177:763–774

Giri DK, Aggarwal BB (1998) Constitutive activation of NF-kappaB causes resistance to apoptosis in human cutaneous T cell lymphoma HuT-78 cells. Autocrine role of tumor necrosis factor and reactive oxygen intermediates. Biol J Chem 273:14008–14014

Goetze S, Xi XP, Kawano Y, Kawano H, Fleck E, Hsueh WA, Law RE (1999) TNF-alpha-induced migration of vascular smooth muscle cells is MAPK dependent. Hypertension 33:183–189

Grell M, Douni E, Wajant H, Lohden M, Clauss M, Maxeiner B, Georgopoulos S, Lesslauer W, Kollias G, Pfizenmaier K et al. (1995) The transmembrane form of tumor necrosis factor is the prime activating ligand of the 80 kDa tumor necrosis factor receptor. Cell 83:793–802

Han SS, Keum YS, Seo HJ, Chun KS, Lee SS, Surh YJ (2001) Capsaicin suppresses phorbol ester-induced activation of NF-kappaB/Rel and AP-1 transcription factors in mouse epidermis. Cancer Lett 164:119–126

Hanauer SB (2004) Efficacy and safety of tumor necrosis factor antagonists in Crohn's disease: overview of randomized clinical studies. Rev Gastroenterol Disord 4 [Suppl 3]:S18–S24

Haridas V, Darnay BG, Natarajan K, Heller R, Aggarwal BB (1998) Over-expression of the p80 TNF receptor leads to TNF-dependent apoptosis, nuclear factor-kappa B activation, and c-Jun kinase activation. Immunol J 160:3152–3162

Hideshima T, Chauhan D, Richardson P, Mitsiades C, Mitsiades N, Hayashi T, Munshi N, Dang L, Castro A, Palombella V, Adams J, Anderson KC (2002) NF-kappa B as a therapeutic target in multiple myeloma. J Biol Chem 277:16639–16647

Henry JR, Rupert KC, Dodd JH, Turchi IJ, Wadsworth SA, Cavender DE, Schafer PH, Siekierka JJ (1998) Potent inhibitors of the MAP kinase p38. Bioorg Med Chem Lett 8:3335–3340

Hernandez R, Teruel T, de Alvaro C, Lorenzo M (2004) Rosiglitazone ameliorates insulin resistance in brown adipocytes of Wistar rats by impairing TNF-alpha induction of p38 and p42/p44 mitogen-activated protein kinases. Diabetologia 47:1615–1624

Hsu SM, Hsu PL (1990) Lack of effect of colony-stimulating factors, interleukins, interferons, and tumor necrosis factor on the growth and differentiation of cultured Reed-Sternberg cells. Comparison with effects of phorbol
ester and retinoic acid. Am Pathol J 136:181–189

Ishii M, Suzuki Y, Takeshita K, Miyao N, Kudo H, Hiraoka R, Nishio K, Sato N,
Naoki K, Aoki T, Yamaguchi K (2004) Inhibition of c-Jun NH2-terminal
kinase activity improves ischemia/reperfusion injury in rat lungs. Immunol
J 172:2569–2577

Johnson GL, Lapadat R (2002) Mitogen-activated protein kinase pathways mediated by ERK, JNK, and p38 protein kinases. Science 298:1911–1912

Kane D, FitzGerald O (2004) Tumor necrosis factor-alpha in psoriasis and
psoriatic arthritis: a clinical, genetic, and histopathologic perspective. Curr
Rheumatol Rep 6:292–298

Kishore N, Sommers C, Mathialagan S, Guzova J, Yao M, Hauser S, Huynh K,
Bonar S, Mielke C, Albee L, Weier R, Graneto M, Hanau C, Perry T,
Tripp CS (2003) A selective IKK-2 inhibitor blocks NF-kappa B-dependent
gene expression in interleukin-1 beta-stimulated synovial fibroblasts. J Biol
Chem 278:32861–32871

Kumar S, Mc PCDonnell, Gum RJ, Hand AT, Lee JC, Young PR (1997) Novel
homologues of CSBP/p38 MAP kinase: activation, substrate specificity and
sensitivity to inhibition by pyridinyl imidazoles. Biochem Biophys Res
Commun 235:533–538

Kumar A, Dhawan S, Hardegen NJ, Aggarwal BB (1998a) Curcumin (diferuloylmethane) inhibition of tumor necrosis factor (TNF)-mediated adhesion
of monocytes to endothelial cells by suppression of cell surface expression
of adhesion molecules and of nuclear factor-kappaB activation. Biochem
Pharmacol 55:775–783

Kumar A, Middleton A, Chambers TC, Mehta KD (1998b) Differential roles
of extracellular signal-regulated kinase-1/2 and p38(MAPK) in interleukin-
1beta- and tumor necrosis factor-alpha-induced low density lipoprotein receptor expression in HepG2 cells. Biol J Chem 273:15742–15748

Kumar A, Takada Y, Boriek AM, Aggarwal BB (2004) Nuclear factor-kappaB:
its role in health and disease. Mol J Med 82:434–448

Kurzrock R, Rosenblum MG, Sherwin SA, Rios A, Talpaz M, Quesada JR, Gut-
terman JU (1985) Pharmacokinetics, single-dose tolerance, and biological
activity of recombinant gamma-interferon in cancer patients. Cancer Res
45:2866–2872

Lee JC, Laydon JT, McDonnell PC, Gallagher TF, Kumar S, Green D, Mc-
Nulty D, Blumenthal MJ, Heys JR, Landvatter SW et al (1994) A protein
kinase involved in the regulation of inflammatory cytokine biosynthesis.
Nature 372:739–746

Lee JH, Slifman NR, Gershon SK, Edwards ET, Schwieterman WD, Siegel JN, Wise RP, Brown SL, Udall JN Jr, Braun MM (2002) Life-threatening histoplasmosis complicating immunotherapy with tumor necrosis factor alpha antagonists infliximab and etanercept. Arthritis Rheum 46:2565–2570

Lukita-Atmadja W, Ito Y, Baker GL, McCuskey RS (2002) Effect of curcuminoids as anti-inflammatory agents on the hepatic microvascular response to endotoxin. Shock 17:399–403

Manna SK, Aggarwal BB (2000) Vesnarinone suppresses TNF-induced activation of NF-kappa B, c-Jun kinase, and apoptosis. Immunol J 164:5815–5825

Manna SK, Sah NK, Newman RA, Cisneros A, Aggarwal BB (2000) Oleandrin suppresses activation of nuclear transcription factor-kappaB, activator protein-1, and c-Jun NH2-terminal kinase. Cancer Res 60:3838–3847

Masson C, Simon V, Hoppe E, Insalaco P, Cisse I, Audran M (2004) Tumor necrosis factor receptor-associated periodic syndrome (TRAPS): definition, semiology, prognosis, pathogenesis, treatment, and place relative to other periodic joint diseases. Joint Bone Spine 71:284–290

Messadi DV, Doung HS, Zhang Q, Kelly AP, Tuan TL, Reichenber E, Le AD (2004) Activation of NFkappaB signal pathways in keloid fibroblasts. Arch Dermatol Res 296:125–133

Mohan N, Edwards ET, Cupps TR, Slifman N, Lee JH, Siegel JN, Braun MM (2004) Leukocytoclastic vasculitis associated with tumor necrosis factor-alpha blocking agents. Rheumatol J 31:1955–1958

Mukhopadhyay A, Suttles J, Stout RD, Aggarwal BB (2001) Genetic deletion of the tumor necrosis factor receptor p60 or p80 abrogates ligand-mediated activation of nuclear factor-kappa B and of mitogen-activated protein kinases in macrophages. Biol J Chem 276:31906–31912

Murakami Y, Shoji M, Hanazawa S, Tanaka S, Fujisawa S (2003) Preventive effect of bis-eugenol, a eugenol ortho dimer, on lipopolysaccharide-stimulated nuclear factor kappa B activation and inflammatory cytokine expression in macrophages. Biochem Pharmacol 66:1061–1066

Murakami-Mori K, Mori S, Nakamura S (1999) p38MAP kinase is a negative regulator for ERK1/2-mediated growth of AIDS-associated Kaposi's sarcoma cells. Biochem Biophys Res Commun 264:676–682

Murata T, Shimada M, Sakakibara S, Yoshino T, Masuda T, Shintani T, Sato H, Koriyama Y, Fukushima K, Nunami N, Yamauchi M, Fuchikami K, Komura H, Watanabe A, Ziegelbauer KB, Bacon KB, Lowinger TB (2004) Synthesis and structure–activity relationships of novel IKK-beta inhibitors. Part 3: orally active anti-inflammatory agents. Bioorg Med Chem Lett 14:4019–4022

Nagata S, Golstein P (1995) The Fas death factor. Science 267:1449–1456

Ozes ON, Mayo LD, Gustin JA, Pfeffer SR, Pfeffer LM, Donner DB (1999) NF-kappaB activation by tumour necrosis factor requires the Akt serine-threonine kinase. Nature 401:82–85

Palanki MS, Erdman PE, Ren M, Suto M, Bennett BL, Manning A, Ransone L, Spooner C, Desai S, Ow A, Totsuka R, Tsao P, Toriumi W (2003) The design and synthesis of novel orally active inhibitors of AP-1 and NF-kappaB mediated transcriptional activation. SAR of in vitro and in vivo studies. Bioorg Med Chem Lett 13:4077–4080

Pastorino JG, Tafani M, Farber JL (1999) Tumor necrosis factor induces phosphorylation and translocation of BAD through a phosphatidylinositide-3-OH kinase-dependent pathway. Biol J Chem 274:19411–19416

Reiley W, Zhang M, Sun SC (2004) Negative regulation of JNK signaling pathway by the tumor suppressor CYLD. Biol J Chem 279:55161–55167

Sandborn WJ, Hanauer S, Loftus EV Jr, Tremaine WJ, Kane S, Cohen R, Hanson K, Johnson T, Schmitt D, Jeche R (2004) An open-label study of the human anti-TNF monoclonal antibody adalimumab in subjects with prior loss of response or intolerance to infliximab for Crohn's disease. Am Gastroente Jrol 99:1984–1989

Shishodia S, Majumdar S, Banerjee S, Aggarwal BB (2003) Ursolic acid inhibits nuclear factor-kappaB activation induced by carcinogenic agents through suppression of IkappaBalpha kinase and p65 phosphorylation: correlation with down-regulation of cyclooxygenase 2, matrix metalloproteinase 9, and cyclin D1. Cancer Res 63:4375–4383

Slifman NR, Gershon SK, Lee JH, Edwards ET, Braun MM (2003) Listeria monocytogenes infection as a complication of treatment with tumor necrosis factor alpha-neutralizing agents. Arthritis Rheum 48:319–324

Steffan RJ, Matelan E, Ashwell MA, Moore WJ, Solvibile WR, Trybulski E, Chadwick CC, Chippari S, Kenney T, Eckert A, Borges-Marcucci L, Keith JC, Xu Z, Mosyak L, Harnish DC (2004) Synthesis and activity of substituted 4-(indazol-3-yl)phenols as pathway-selective estrogen receptor ligands useful in the treatment of rheumatoid arthritis. J Med Chem 47:6435–6438

Sugarman BJ, Aggarwal BB, Hass PE, Figari IS, Palladino MA Jr, Shepard HM (1985) Recombinant human tumor necrosis factor-alpha: effects on proliferation of normal and transformed cells in vitro. Science 230:943–945

Sugarman BJ, Lewis GD, Eessalu TE, Aggarwal BB, Shepard HM (1987) Effects of growth factors on the antiproliferative activity of tumor necrosis factors. Cancer Res 47:780–786

Takada Y, Aggarwal BB (2003a) Betulinic acid suppresses carcinogen-induced NF-kappa B activation through inhibition of I kappa B alpha kinase and p65 phosphorylation: abrogation of cyclooxygenase-2 and matrix metalloprotease-9. Immunol J 171:3278–3286

Takada Y, Aggarwal BB (2003b) Genetic deletion of the tumor necrosis factor receptor p60 or p80 sensitizes macrophages to lipopolysaccharide-induced nuclear factor-kappa B, mitogen-activated protein kinases, and apoptosis. Biol J Chem 278:23390–23397

Takada Y, Aggarwal BB (2004) Evidence that genetic deletion of the TNF receptor p60 or p80 in macrophages modulates RANKL-induced signaling. Blood 104:4113–4121

Tartaglia LA, Pennica D, Goeddel DV (1993) Ligand passing: the 75-kDa tumor necrosis factor (TNF) receptor recruits TNF for signaling by the 55-kDa TNF receptor. Biol J Chem 268:18542–18548

Torii A, Miyake M, Morishita M, Ito K, Torii S, Sakamoto T (2004) Vitamin A reduces lung granulomatous inflammation with eosinophilic and neutrophilic infiltration in Sephadex-treated rats. Eur Pharmacol J 497:335–342

Tran SE, Holmstrom TH, Ahonen M, Kahari VM, Eriksson JE (2001) MAPK/ERK overrides the apoptotic signaling from Fas, TNF, TRAIL receptors. Biol J Chem 276:16484–16490

Tucker SJ, Rae C, Littlejohn AF, Paul A, MacEwan DJ (2004) Switching leukemia cell phenotype between life and death. Proc Natl Acad Sci U S A 101:12940–12945

Vanden Berghe W, Plaisance S, Boone E, De Bosscher K, Schmitz ML, Fiers W, Haegeman G (1998) p38 and extracellular signal-regulated kinase mitogen-activated protein kinase pathways are required for nuclear factor-kappaB p65 transactivation mediated by tumor necrosis factor. Biol J Chem 273:3285–3290

Wallis RS, Vinhas SA, Johnson JL, Ribeiro FC, Palaci M, Peres RL, Sa RT, Dietze R, Chiunda A, Eisenach K, Ellner JJ (2003) Whole blood bactericidal activity during treatment of pulmonary tuberculosis. Infect J Dis 187:270–278

Wolfe F, Michaud K (2004) Lymphoma in rheumatoid arthritis: the effect of methotrexate and anti-tumor necrosis factor therapy in 18,572 patients. Arthritis Rheum 50:1740–1751

Yang L, Dan HC, Sun M, Liu Q, Sun XM, Feldman RI, Hamilton AD, Polokoff M, Nicosia SV, Herlyn M, Sebti SM, Cheng JQ (2004) Akt/protein kinase B signaling inhibitor-2, a selective small molecule inhibitor of Akt signaling with antitumor activity in cancer cells overexpressing Akt. Cancer Res 64:4394–4399

11 IL-1 Family Members in Inflammatory Skin Disease

J. Sims, J. Towne, H. Blumberg

References 190

Abstract. The cytokines IL-1α and IL-1β have long been known to play a profound role in inflammation, and in the past decade another cytokine, IL-18 (originally known as IGIF), has also been realized to be an IL-1 family member and to possess significant inflammatory activity. Half a dozen additional members of the IL-1 family have been identified in recent years, and given their relatedness to IL-1 and IL-18, it is tempting to speculate that they too might possess inflammatory potential. We have demonstrated that certain of these cytokines can activate MAP kinases and the pathway leading to NFκB, via known IL-1R family members. Moreover, when overexpressed in skin, they are capable of causing an inflammatory skin condition resembling that seen in human disease.

The cytokines IL-1α and IL-1β have long been known to play a profound role in inflammation, acting on virtually every organ system and cell type in the body (Dinarello 2002) to induce other cytokines, chemokines, adhesion molecules, and mediators such as NO and prostaglandins. Nine years ago, a new cytokine, originally known as IGIF and now termed IL-18, was cloned and soon realized to be related by descent to IL-1 (Okamura et al. 1995; Bazan et al. 1996). IL-18 also acts in pro-inflammatory fashion, primarily by its ability (under normal conditions) to promote expansion and differentiation of the Th1 helper T cell population and to induce the Th1 cytokine, interferon gamma (IFN-γ). Since that time, the IL-1 family has continued to expand so that it is now comprised of ten members (Dunn et al. 2001; Sims et al. 2001). The grouping of these ten genes into a family is based on conservation of sequence, of three-dimensional structure, and of gene organization and location (Smith et al. 2000; Taylor et al. 2002; Nicklin et al. 2002; Dunn et al. 2003).

IL-1 exerts its activities by binding to the type I IL-1 receptor and subsequently recruiting a homologous molecule, the IL-1 receptor accessory protein (AcP) (Sims 2002). It is the heterodimeric receptor complex that is competent for signaling. IL-18 acts in similar fashion, binding to IL-18Rα and subsequently recruiting an accessory chain, IL-18Rβ, which initiates biological activities (Sims 2002). The polypeptides comprising both chains of the IL-1 and IL-18 receptor complexes are members of the IL-1 receptor family, which was originally defined by proteins comprised of three immunoglobulin domains in the extracellular portion, a single transmembrane domain, and a cytoplasmic portion, which contains a TIR (toll interleukin-1 receptor) domain. As the IL-1 ligand family has expanded, so has the IL-1 receptor family, to include nine members (Born et al. 2000). One of the newer members has only one Ig domain, and several have extra amino acids C-terminal to the TIR domain, but the essential family attributes remain.

The similarities between IL-1, IL-18, and their receptors, on the one hand, and the new IL-1 and IL-1R family members, on the other hand, readily leads to speculation that the latter may, like their previously characterized cousins, play roles in inflammation. Indeed, a report in 2001 (Debets) claimed that one of the new ligands, IL-1F9, was able to induce NFκB activity in Jurkat cells as long as the cells had been transfected with a formerly orphan IL-1R family member, IL-1R rp2. We were able to confirm this result, and extend it to demonstrate that not only IL-1F9, but also IL-1F6 and IL-1F8, possessed the ability to activate the pathway leading to NFκB (Towne et al. 2004). All three ligands do so not only in Jurkat cells, but in many human and mouse cell lines, as long as the cells express IL-1R rp2, and can induce activation of the MAP kinases Erk and JNK in addition to NFκB. Antibody-blocking and transfection experiments suggest that both AcP (the same AcP that plays a role in IL-1 signaling) and IL-1R rp2 are required for these responses. There are two unusual aspects to the activity of IL-1F6, F8, and F9, however, which remain unresolved. One is the high ligand concentration required for signaling. The cell lines assayed to date do not respond to these ligands at concentrations less than 10 ng/ml, and concentrations of 1–5 ug/ml are required in order to obtain a maximal response. Both of these values are several orders of magnitude higher than is seen with IL-1, for example, and about one order of magnitude higher than demonstrated by IL-18. Second, we have not been able to detect binding of IL-1F6, F8, or F9 to either IL-1R rp2 or AcP, by a variety of techniques including surface plasmon resonance. Whether these results suggest the existence of an additional receptor subunit, not expressed on our assay cell lines and not required for signaling but providing extra affinity for the ligand, or some other explanation, is not currently clear.

In vivo, IL-1α has been shown to be capable of inducing an inflammatory skin condition when expressed under control of the keratin-14 promoter at levels approximately ten times greater than that seen following the potent inflammatory stimulus LPS (Groves et al. 1995). However, when expressed constitutively at levels similar to those induced by LPS, which are still very high, the inflammatory skin phenotype is only occasionally seen. Skin-specific expression of IL-18 also results in an inflammatory condition, although one that is slow to develop (Konishi et al. 2002). Overexpression of IL-18 in the skin also exacerbates dermatitis elicited by irritants or contact hypersensitivity (Kawase et al. 2003). Interestingly, skin-specific expression of caspase 1, the protease responsible for cleaving the inactive precursors of both IL-18 and IL-1β to their active forms, leads to a chronic erosive dermatitis developing at about 8 weeks of age and persisting through the life of the animal (Yamanaka et al. 2000). Double-mutant animals transgenic for keratin14-expressed caspase 1 but deficient in IL-18 failed to develop the dermatitis. Double-mutant animals transgenic for keratin14-expressed caspase 1 but deficient in IL-1 did develop the dermatitis but with a delayed time course. These results suggest that the pathological process leading to dermatitis in the caspase-1 transgenic mice requires IL-18, and is accelerated by IL-1 (Konishi et al. 2002).

In order to determine the in vivo effects of the newer IL-1 family members, we generated transgenic mice expressing either IL-1F6, IL-1F8, or IL-1R rp2 from the keratin 14 promoter. This promoter drives expression predominantly in the basal epithelium of the skin, although there is also some expression in the thymus, tongue, and forestomach of the mouse. Mice overexpressing any of these three genes under control of the K14 promoter develop an inflammatory skin disease, which histologically bears some resemblance to human psoriatic skin. Consistent with this finding, IL-1F6 and F8 can be seen by in situ hybridization to be present at elevated levels in human psoriatic skin, compared to skin from nondiseased individuals. Future studies will explore further the role of the novel IL-1 family members in human dermatological disease.

References

Bazan JF, Timans JC, Kastelein RA (1996) A newly defined interleukin-1? Nature 379:591

Born TL, Smith DE, Garka KE, Renshaw BR, Bertles JS, Sims JE (2000) Identification and characterization of two members of a novel class of the interleukin-1 receptor (IL-1R) family. Delineation of a new class of IL-1R-related proteins based on signaling. J Biol Chem 275:29946–29954

Dinarello CA (2002) The IL-1 family and inflammatory diseases. Clin Exp Rheumatol 20[Suppl 27]:S1–S1.

Dunn EF, Gay NJ, Bristow AF, Gearing DP, O'Neill LA, Pei XY (2003) High-resolution structure of murine interleukin 1 homologue IL-1F5 reveals unique loop conformations for receptor binding specificity. Biochemistry 42:10938–10944

Dunn E, Sims JE, Nicklin MJ, O'Neill LA (2001) Annotating genes with potential roles in the immune system: six new members of the IL-1 family. Trends Immunol 22:533–536

Groves RW, Mizutani H, Kieffer JD, Kupper TS (1995) Inflammatory skin disease in transgenic mice that express high levels of interleukin 1 alpha in basal epidermis. Proc Natl Acad Sci U S A 92:11874–11878

Kawase Y, Hoshino T, Yokota K, Kuzuhara A, Kirii Y, Nishiwaki E, Maeda Y, Takeda J, Okamoto M, Kato S, Imaizumi T, Aizawa H, Yoshino K (2003) Exacerbated and prolonged allergic and non-allergic inflammatory cutaneous reaction in mice with targeted interleukin-18 expression in the skin. J Invest Dermatol 121:502–509

Konishi H, Tsutsui H, Murakami T, Yumikura-Futatsugi S, Yamanaka K, Tanaka M, Iwakura Y, Suzuki N, Takeda K, Akira S, Nakanishi K, Mizutani H (2002) L-18 contributes to the spontaneous development of atopic dermatitis-like inflammatory skin lesion independently of IgE/stat6 under specific pathogen-free conditions. Proc Natl Acad Sci U S A 99:11340–11345

Nicklin MJ, Barton JL, Nguyen M, FitzGerald MG, Duff GW, Kornman K (2002) A sequence-based map of the nine genes of the human interleukin-1 cluster. Genomics 79:718–725

Okamura H, Tsutsi H, Komatsu T, Yutsudo M, Hakura A, Tanimoto T, Torigoe K, Okura T, Nukada Y, Hattori K et al (1995) Cloning of a new cytokine that induces IFN-gamma production by T cells Nature 378(6552):88–91

Sims JE (2002) IL-1 and IL-18 receptors, and their extended family. Curr Opin Immunol 14:117–22

Sims JE, Nicklin MJ, Bazan JF, Barton JL, Busfield SJ, Ford JE, Kastelein RA, Kumar S, Lin H, Mulero JJ, Pan J, Pan Y, Smith DE, Young PR (2001) A new nomenclature for IL-1-family genes. Trends Immunol 22 :536–537

Smith DE, Renshaw BR, Ketchem RR, Kubin M, Garka KE, Sims JE (2000) Four new members expand the interleukin-1 superfamily. J Biol Chem 275:1169–1175

Taylor SL, Renshaw BR, Garka KE, Smith DE, Sims JE (2002) Genomic organization of the interleukin-1 locus. Genomics 79:726–733

Towne JE, Garka KE, Renshaw BR, Virca GD, Sims JE (2004) Interleukin (IL)-1F6, IL-1F8, and IL-1F9 signal through IL-1Rrp2 and IL-1RAcP to activate the pathway leading to NF-kappaB and MAPKs. J Biol Chem 279:13677–13688

Yamanaka K, Tanaka M, Tsutsui H, Kupper TS, Asahi K, Okamura H, Nakanishi K, Suzuki M, Kayagaki N, Black RA, Miller DK, Nakashima K, Shimizu M, Mizutani H (2000) Skin-specific caspase-1-transgenic mice show cutaneous apoptosis and pre-endotoxin shock condition with a high serum level of IL-18. J Immunol 165:997–1003

12 Regulatory T Cells in Psoriasis

M.H. Kagen, T.S. McCormick, K.D. Cooper

12.1   Introduction 194
12.1.1 Psoriasis as a T Cell-Mediated Autoimmune Disease 194
12.2   T Cell Activation 196
12.3   T Regulatory Cells 197
12.3.1 Identifying Treg Cells 197
12.4   T Regulatory Cells in Psoriasis 199
12.4.1 Numbers 200
12.4.2 Activity 200
12.4.3 Tissue 203
12.4.4 Measuring Treg Cells in Skin 203
12.5   Therapy and the Treg Population: The Example of Alefacept . . . 205
12.6   Conclusions and Implications 206
References 207

Abstract. Psoriasis is a chronic autoimmune disease in which T lymphocytes are thought to be central in the pathogenesis. Recently, a T cell subset population was identified, whose role is to suppress inflammatory responses triggered by T effector cells. T cells in this new population are referred to as T regulatory cells. We studied their number and activity in psoriatic lesions and found that they are both numerically and functionally deficient in their ability to suppress the abnormally persistent psoriatic immune response. This deficiency may shed more light on the complex pathophysiology of psoriasis.

12.1 Introduction

Psoriasis is a common cutaneous disease affecting as many as 2% of the population in the world. Though there have been differing viewpoints regarding the pathogenesis of psoriasis over the preceding decades, common consensus holds that it is an autoimmune disease, and likely the most common autoimmune disease. Though an etiology of psoriasis still has not been determined, activated T cells appear to be central in promoting both disease onset and maintenance. This determination arrives from several observations:

1. T cells are seen to appear in nonlesional skin before there is evidence of keratinocyte proliferation; these T cells are usually of the memory (CD45RO$^+$) subtype.
2. Therapy specifically directed against T cells may be remittive in disease.
3. T cells are critical to maintain lesions.

Why T cells remain in a persistently activated state and target normal biologic structure, i.e., keratinocytes, remains unclear, though we can postulate that any of several factors may play a role, perhaps in concert. There may be an overactivity of antigen presentation by antigen presenting cells (APCs) which interact with naïve T cells through receptor pairing and interaction, such as LFA3-ICAM-1, B-7-CD28, etc. APCs may be elaborating stimulatory cytokines such as IL-12, IL-23 TNF-α, or IL-15, activating T cells and driving a differentiation toward the type I or pro-inflammatory phenotype. There may be an abnormal basement membrane zone in prelesional psoriatic skin that is hyperstimulatory. Furthermore, we may postulate that in individuals with psoriasis, there is an intrinsic defect in the ability of elements in the immune system whose function it is to restrain and terminate brisk responses, therefore leading to a persistent and chronically abnormal immune response. These elements of the immune system include what are now recognized to be T regulatory (Treg) cells.

12.1.1 Psoriasis as a T Cell-Mediated Autoimmune Disease

Evidence of psoriasis as a T cell-mediated autoimmune disease is fairly recent, as earlier research had focused on the epidermal hyperplasia as the more primary phenomenon. However, even before antibodies were available to phenotype different types of leukocytes, it was observed that the dermis of psoriasis vulgaris lesions contained numerous mononuclear cells, and that these cells appeared in early lesions before obvious epidermal changes were apparent (Braun-Falco and Schmoekel 1977). We now recognize these cells as T cells, and in fact, the earliest appearing T cells are CD45RO$^+$ memory T cells (Vissers et al 2004); the CD45RO memory (effector) population actually undergoes proliferation within the psoriatic lesion (Morganroth et al. 1991). Phenotypic analysis of infiltrating T cells reveals that a majority of the infiltrate in the dermis is CD4$^+$, whereas most of the epidermal T cells are CD8$^+$ (cyototoxic T cells) (Krueger 2002). By both analysis of surface markers and measures of cytokines produced, it is apparent that the majority of these infiltrating T cells are type I, or prinflammatory and produce cytokines such IFN-γ, TNF-α, IL-8 and IL-2 as opposed to more anti-inflammatory or humoral cytokines, such as IL-4, IL-5, IL-10, or IL-13. (Szabo et al. 1998; Austin et al. 1999; Lew et al. 2004) Over time, both through serendipity and by design, it has been observed that agents that either inhibit T cell activity (such as steroids or cyclosporine), or that were directly toxic to these cells (DAB$_{389}$IL-2, UV light) were associated with clearance of psoriasis. As such, it is now the current consensus that pathogenic T effector cells mediate both the initiation and maintenance of chronic psoriatic plaques.

Several have noticed that the basement membrane of the epidermis in patients with psoriasis is ultrastructrally different than that of phenotypically normal individuals (Heng et al. 1986). Concordantly, uninvolved and involved skin of patients with psoriasis exhibits increased basement membrane zone expression of the EDA splice variant of fibronectin, a form that is associated with hyperplasia, such as in wounds, fetal tissue, and cancer (Bata-Csorgo et al. 1998; Ting et al. 2000). Functionally, psoriatic basal keratinocytes containing resting stem cells exposed to fibronectin undergo more rapid induction of cell cycle entry and cell spreading if the cells are derived from psoriatic uninvolved vs normal skin, particularly in combination with psoriatic T cell lymphokines (Bata Csorgo et al. 1995; Chen et al. 2001; Szeli et al. 2004). This is, it appears, central in the pathogenesis of psoriasis is the interplay of activated T cells, antigen-presenting cells, and an abnormally primed hyperplastic

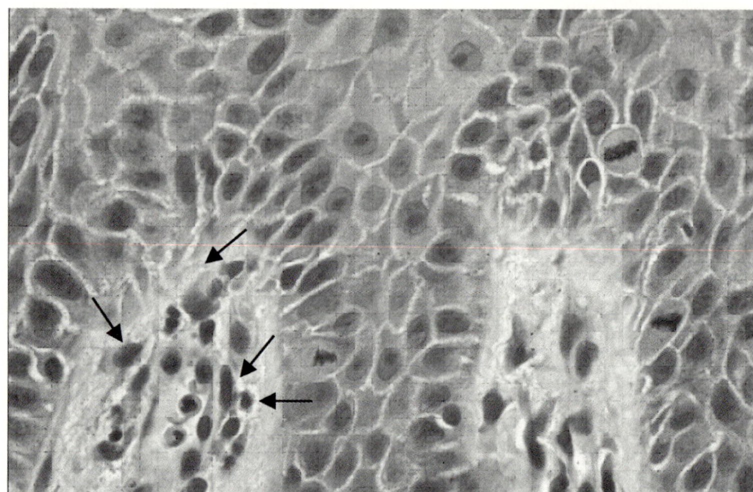

Fig. 1. Critical area of interaction between perivascular T cells, macrophages, dendritic cells, and basal keratinocytes

basal keratinocyte layer (Fig. 1). As we will subsequently discuss, there is also likely a lack of an internal regulatory check on these events.

12.2   T Cell Activation

T cells that appear in psoriatic lesions are activated, that is, they presumably have encountered an antigen presented by an antigen-presenting cell (APC) and have become primed to proliferate and produce proinflammatory cytokines and recruit other inflammatory cell types to the target area. T cell activation requires two distinct molecular steps. First, the T cell must come into direct apposition with an APC by way of the interaction of the T cell receptor (TCR) on the T cell with the MHC receptor (bearing the antigen) on the APC. Additionally, there must also be a number of secondary interactions between receptor pairs on the T cell and APC, which are called costimulatory pairs. This additional costimulation is necessary for T cell activation. Examples of such costimulatory pairs are CD2, LFA-1, and CD28 on the T cell paired with LFA-3, ICAM-1, and B7 on the APC, respectively. Naïve T cells with appropriate specificity for antigens or autoantigens encountered in psoriatic skin are thus continually recruited to activate, mature, and proliferate as memory/effector pathogenic T cells. Because psoriasis is dominated by type 1 T cells producing IFN-γ, the emerging T cells are continuously influenced to mature into type 1 T cells, under the influence of IFN-γ-induced T-bet, and APC-derived IL-12 and/or IL-23. Confirmation of the activity of memory/effector Th1 maturation is provided by the ability of alefacept (LFA-3-FcIgG dimer) and anti IL-12 /23 p40 to clear psoriasis (Ellis and Krueger 2001).

12.3 T Regulatory Cells

Constraints on T memory/effector cell expansion occur in normal immune responses, limiting tissue damage, autoimmunity, and lymphoma development. Thus, in the normal physiologic state, immune responses are designed to be transient, that is, they are shut off once they have dealt with whatever agent triggered their activation. It has been known that Th1 and Th2 cytokines and transcription factors are mutually inhibitory, and either type of cytokine may terminate its counterpart's response. Also recently identified is another naturally occurring cell type, whose main function is to inhibit T cell activation and proliferation (an inflammatory immune response), the T regulatory (Treg) cell.

12.3.1 Identifying Treg Cells

Treg cells are $CD4^+$ and constitutively express CD25, the α chain of the IL-2 receptor. In this, they may sometimes be difficult to distinguish from chronically activated T cells, which also express CD25. However, as will be seen subsequently, when analyzed using flow cytometry, they express a distinctly higher fluorescence intensity than activated or Tresponder (Tresp) cells. Also, unlike Tresp cells, they are $CTLA-4^+$ (an anti-inflammatory marker) and express the glucocorticoid induced tumor necrosis factor receptor, GITR. Additionally, Treg cells express very high levels of foxp3, a transcription factor thought to program the development and function of this subset of cells (Ramsdell 2003; Akbar et al. 2003; Annacker et al. 2001; Fehevari and Sakaguchi 2004).

Treg cells constitute about 10% of murine CD4$^+$ cells. Murine Treg cells have been shown to inhibit autoimmune gastritis after thymectomy. They are anergic when stimulated with monoclonal anti-CD3 antibodies, but proliferate with the addition of IL-2. They show remarkable suppressive activity both in vitro and in vivo. After TCR-mediated stimulation, Treg cells suppress the activation and proliferation of other CD4$^+$ and CD8$^+$ T cells in an antigen nonspecific manner, via a mechanism that

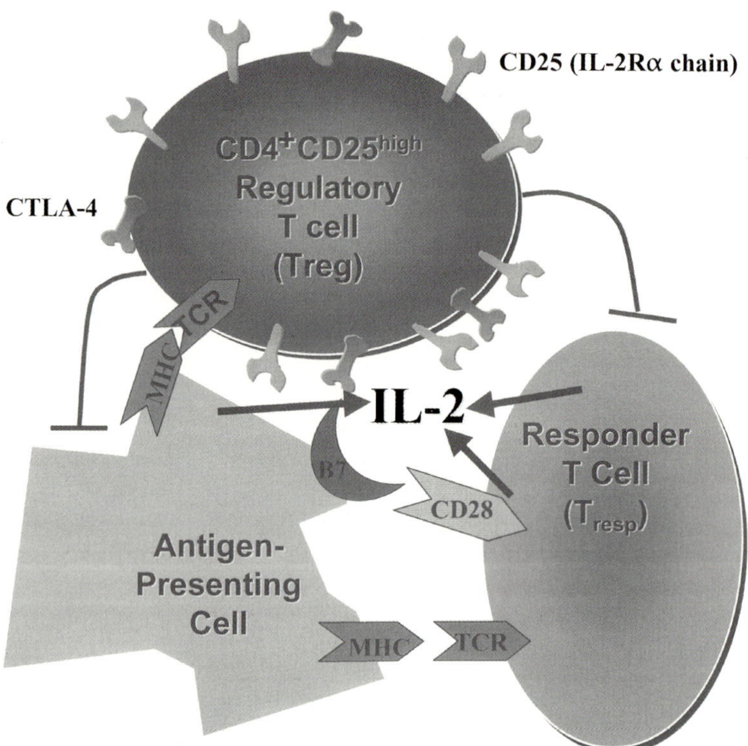

Fig. 2. A Treg cell coming into apposition with an interacting APC–Tresp pair through ligation of the TCR on the Treg cell with an MHC class molecule on the APC. Both the APC and the Tresp cell secrete IL-2, which by binding to CD25 expressed on the Treg cell surface, further enhances the Treg mediated immunosuppression and may induce the Treg cell to proliferate

requires cell-to-cell contact and in most systems is independent of the production of immunosuppressive cytokines. In humans, Treg cells appear to be similar to those seen in mice, both in number and function. They fail to proliferate in response to polyclonal stimulation and are able to suppress T cell responses in a cytokine-independent, but contact-dependent manner (Baecher-Allen et al. 2001, 2004). Antigen specificity may be derived from their ability to suppress only via cell contact during the immunologic synapse between the APC, the responding antigen-specific T cell, and the Treg cell (Fig. 2). IL-2 secreted into the micromilieu by the bidirectional activation of the APC and T responder cell, in conjunction with B7-CTLA4/CD28 and TCR activation of the Treg, serves to activate the Treg to suppress the responding T cell, and likely serves to locally expand the Treg cell population as well.

Treg cells serve an important function in inducing peripheral tolerance. They have been of crucial importance in the organ transplant setting and are becoming of increasing interest in understanding autoimmunity, since it is likely that successful Treg function acts as a check against this phenomenon (Hori et al. 2003; Chatenoud et al. 2001; Roncarolo and Levings 2000; Shevach 2004). As such, it is of interest to investigate whether there are imbalances in the Treg population in an autoimmune disease such as psoriasis, and what effect therapy may have on this population.

12.4   T Regulatory Cells in Psoriasis

As mentioned, Tregcells are identified either in the circulation or tissue by characteristic surface markers. They are $CD4^+$ and $CD25^+$. Chronically activated T cells are also $CD25^+$, although they have a distinctly lower mean fluorescence intensity (MFI) than Treg cells as measured by flow cytometry and are referred to as a $CD25^{low}$ T cells. Conversely Treg cells are referred to as $CD25^{high}$ or $CD25^{mid}$.

12.4.1 Numbers

Peripheral Blood To assay whether there was a numeric difference in Treg cells between patients with or without psoriasis, PBMCs were collected from six patients with psoriasis and eight normal volunteers and analyzed by flow cytometry. There was no statistically significant difference seen between the proportions of $CD4^+CD25^{high/mid}$ T cells, either in the psoriatic patients or normal controls.

12.4.2 Activity

As mentioned, Treg cells are able to exert an inhibitory stimulus on Teffector cells and are thus able to control and to terminate inflammatory responses in the normal setting. We determined to demonstrate this phenomenon by quantitatively measuring Teffector proliferation as assayed by ^3H thymidine uptake in response to an antigen with or without the presence of Treg cells in both normal volunteers and psoriatic patients. $CD4^+$ Treg cells were selected by anti-CD25 beads, which resulted in a preparation of $CD25^{high}$ and $CD25^{mid}$ cells. These were mixed at an equal ratio and cultured with allogeneic APCs.

We found that Treg cells did indeed inhibit the ability of Teffector cells to proliferate, although the effect was markedly different in the normal volunteers where inhibition of proliferation was virtually complete as compared to the psoriatic patients, where the inhibition was partial. In contrast to normal $CD4^+CD25^{high}$ Treg cell, that inhibited $CD24^+CD25^-$ Tresp cells by an average of 87.8% at a 1:1 ratio, the inhibitory capacity of psoriatic Treg cells was significantly decreased, (60.6%, $p = 0.0001$), similar to what has been observed for both multiple sclerosis and autoimmune polyglandular syndrome type II (Viglietta et al. 2004; Kriegel et al. 2004) (Fig. 3). This was confirmed using anti-CD3 as well.

In a further effort to quantify the differences between the normal and psoriatic states, the numbers of Treg cells necessary to achieve a specified level of inhibition of proliferation was compared. In the patients with psoriasis, an eightfold higher concentration of Treg:Teffector cells was necessary to reduce proliferation by 50% in the presence of allogeneic antigen stimulation. In normal volunteers, 50% inhibition of

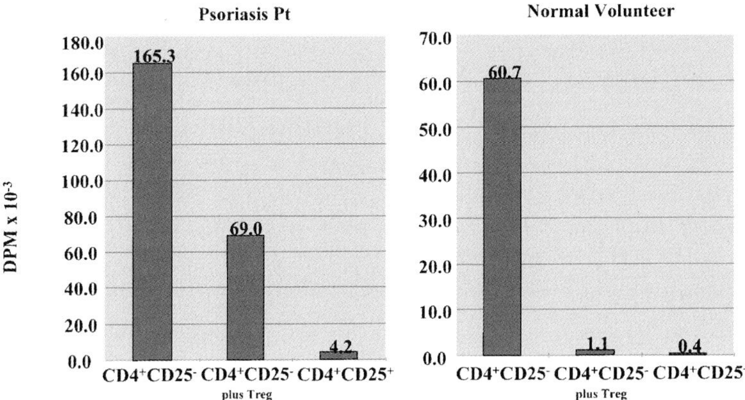

Fig. 3. ³H thymidine uptake in Teffector cells as a marker of proliferation without and then with the addition of Treg cells in the presence of a stimulating antigen. Note the complete inhibition of proliferation in the normal volunteer (*right*) as compared to the partial inhibition in the psoriatic patient (*left*)

proliferation was reached at a Teg:Tresp ratio of 1:16, whereas in the psoriatic patients, this degree of inhibition was not reached until the Treg:Tresp ratio had reached 1:2.

Furthermore, the ability of Treg cells collected from psoriatic patients and normal volunteers to inhibit proliferation was compared to one using two systems to elicit proliferation; using an antigen-specific model, as in the experiment discussed earlier, as well as through polyclonal stimulation of T cell receptor using an antibody to CD3. In both situations, using a Treg:Tresp ratio of 1:1, the psoriatic Treg cells were unable to inhibit proliferation of Tresp cells to anywhere near the degree to that of normal Treg cells, with a difference in inhibition of about 90% inhibition (normal volunteers) to 60% inhibition (psoriatic patients) ($p = 0.0001$). Employing a polyclonal stimulation model did not appreciably alter the results, with a difference of about 90% inhibition in normal volunteers to about 70% inhibition in the patients with psoriasis ($p = 0.029$) (Sugiyama et al. 2005).

These data are quite provocative given that much of the discussion in the last decade centering on the relatively new emphasis on the immunopathogenesis of psoriasis has focused on the abnormally and persistently activated Teffector cells as well as a still fruitless quest for the putative psoriatic antigen(s). The possibility that a defect in the immune system leading to psoriasis might rest with T regulatory cells that are unable to turn off what might otherwise be a normal immune response has not previously been actively pursued.

To further investigate this hypothesis and to determine whether that the abnormality indeed rests with psoriatic Treg cells rather than the more excitable Tresp cells, the following criss-cross experiment was designed and implemented: Treg cells from psoriatic patients were incubated with Tresp cells, APCs, and antigen from normal volunteers, and Treg cells

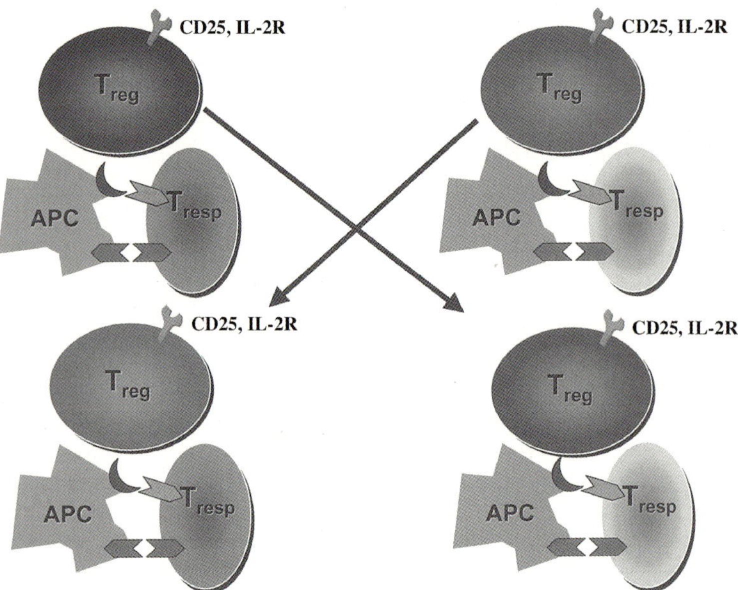

Fig. 4. Normal Treg cells (*blue*) are paired with APCs and Tresp cells from psoriatic patients, while psoriatic Treg cells (*red*) are paired with APCs and Tresp cells from normal patients. Hypothetically, the psoriatic Treg cells should not be able inhibit proliferation of normal Tresp cells as well as the normal Treg cells will inhibit the proliferation of the psoriatic Tresps

from normal volunteers were incubated with Tresp cells, APCs, and antigen from psoriatic patients (Fig. 4). If psoriatic Treg cells are indeed functionally deficient as compared with normal Treg cells, they should not be able to produce the same degree of inhibition that normal Treg cells could.

Interestingly, the psoriatic Treg cells were not as effective in controlling proliferation of Tresp cells, either in the case of psoriatic APCs and Tresp cells or with normal APCs and Tresp cells. Normal Treg cells were more effective at controlling inhibition of both populations. In two separate experiments, using cells from four individuals, at an equal ratio, normal Treg cells suppressed psoriatic Tresp proliferation by an average of 75.2% (compared with with 88.7% inhibition on autologous normal Tresp cells), whereas psoriatic Treg cells were able to suppress normal Treso cell proliferation by only 34.4% (52.6% on autologous Tresp cells) ($p = 0.009$ and $p = 0.011$, respectively). Thus, psoriatic Treg cells are clearly seen to be less potent than normal Treg cells in their ability to inhibit proliferation.

12.4.3 Tissue

Previous discussion and analysis has focused on circulating cells in the periphery. Of potentially more interest and relevance is the cellular in milieu of the skin itself, the affected organ system in psoriasis. A heretofore unaddressed question is whether or not there are Treg cells in the skin and if so, in what numbers are they present and what functionality might they have?

12.4.4 Measuring Treg Cells in Skin

Treg cells were isolated in skin from keratome shave biopsies from patients with psoriasis and identified as such by being highly $CD25^+$ and $CTLA\text{-}4^+$, both being markers that are constitutively expressed by and highly associated with Treg cells. To avoid including activated cells in this group, which may also express both markers, the cells were rested for 48 h before analysis. Tresp cells also upregulate expression of both membrane and cytoplasmic CTLA4 after activation for 24–48 h, but quickly downregulate it thereafter. As such, after a 48-h rest period, the chance of including activated Tresp cells in this group is minimal. As a further check against including Tresp cells in this group, CD69, a marker of early activation was measured and found to be negative in over 99% of the cells. Since the pathogenic events in psoriasis happen in the dermis with the interaction between pathogenic T cells and dividing keratinocytes at the basal layer of the epidermis, we will focus our analysis on the dermal Treg cells. As shown in Fig. 5, Treg cells comprised 14.3% of dermal CD4$^+$ cells.

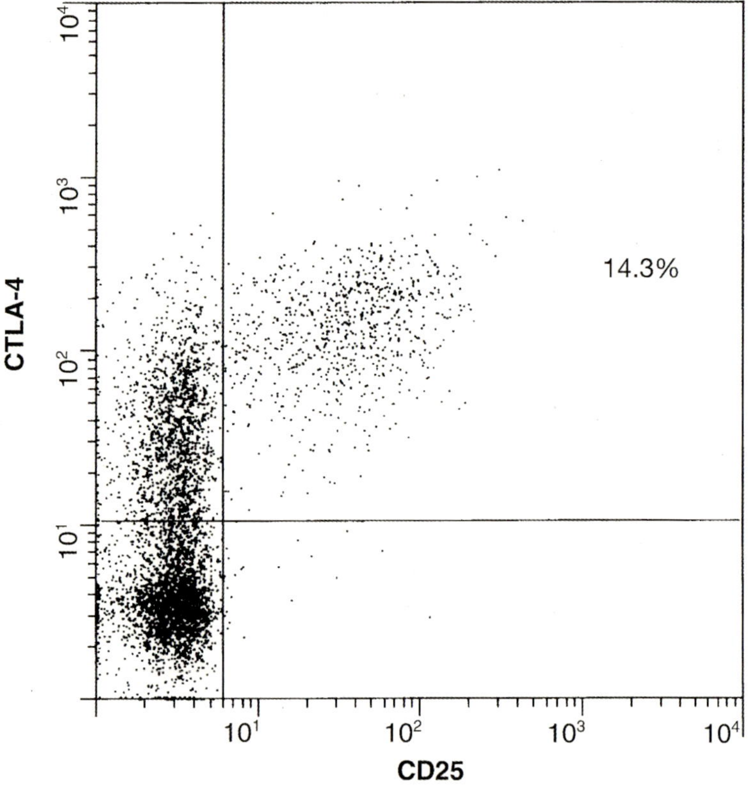

Fig. 5. Percentage of 48-h-rested dermal CD4$^+$ cells that are CTLA-4$^+$ and CD25$^+$, markers of Treg cells. CD25 is on the *x-axis* and CTLA-4 on the *y-axis*

Knowing the approximate the number of Treg cells in the dermis, it was important to know what kind of regulatory or inhibitory function this number of cells could exert in vivo. As such, dermal-derived psoriatic Treg cells were combined with psoriatic Tresp cells and APCs at various Treg:Tresp cell ratios. At a ratio of Treg:Tresp cells of 1:6, near their physiologic concentration, percent inhibition of proliferation was low, at only approximately 20%. Significant inhibition of proliferation was not seen until a Treg:Tresp ratio of 1:3 was reached, which is a concentration more than twice what we determined to be occurring in lesional psoriasis skin.

12.5 Therapy and the Treg Population: The Example of Alefacept

We further wondered whether immunologic therapy that results in long-term remission involves restoring the balance of Treg cells to pathogenic memory/effector cells.

Alefacept (LFA3-TIP) is a bivalent recombinant fusion protein composed of the first extracellular domain of lymphocyte function-associated antigen 3 (LFA-3) fused to the hinge C_H2 and C_H3 domains of human IgG_1. Developed to treat psoriasis, it has a dual mechanism of action. The LFA-3 portion of alefacept binds to CD2 receptors on T cells, blocking their interaction with LFA3 on APCs, and preventing T cell activation. The I_gG1 portion binds to $F_{c\gamma}R$ receptors on natural killer (NK) cells, causing T cell apoptosis via release of granzymes by the NK cells. This apoptotic action preferentially affects memory ($CD45RO^+$) T cells, which become the pathogenic effector cells in psoriasis, presumably because of the higher CD2 expression in these cells (Krueger 2004). Depletion of CD 45 RO^+ T cells is seen in responding patients during the course of alefacept therapy, whereas a similar decrease is not observed in naïve T cells, thus preserving the ability to maintain a functioning immune system and to recognize and counter other foreign challenges.

Alefacept has been reported to cause prolonged disease-free remission in certain patients who responded to treatment. It is relevant to ask whether certain therapies associated with prolonged clearance of disease alter the Teffector:Treg cell ratio in favor of immunosuppression.

As an example, we have measured this ratio in one patient who responded to alefacept therapy and had a prolonged disease-free remission. In this patient, over a 12-week period, the Teffector:Treg cell ratio decreased markedly from approximately 8 to close to 0. Thus, it is possible that repetitive and selective depletion of emerging memory/effector T cells in psoriasis using alefacept, resulted in a diminution in the pathogenic memory/effector population with relative sparing of Treg cells. The diminution in pathogenic T cells may have been sufficient such that the naturally occurring Treg cells were able to be present in an effective ratio that was able to maintain the remission without additional therapy.

12.6 Conclusions and Implications

We have seen that Treg cells, while present in equal amounts in the circulation between people with or without psoriasis, are functionally deficient in preventing T cell activation and responsiveness in psoriasis. The criss-cross experiment illustrated in Fig. 4 established that the deficiency regulation lay in the psoriatic Treg themselves, rather than solely being a factor relating to hyper excitable psoriatic Tresp cells. In addition, in the target tissue, the skin, Treg cells are not present in numbers great enough to maintain adequate control of immune responses. These important observations shed additional light on the complicated and inadequately understood pathogenesis of psoriasis, and suggest that innate deficiencies in immune regulation may play a part in permitting the abnormally chronic immune response clinically seen in the disease. Changes in the number of Treg cells during various therapies may also help explain why some (methotrexate, cyclosporine) produce very short periods of clearance, while others (Alefacept, UVB) are capable of producing longer-term remissions. Of note, UVB has been reported to induce Treg cell production. (Aubin and Mousson 2004) Also of interest is the potential to vaccinate with Treg cells to prevent autoimmunity. (Bluestone and Tang 2004)

Insights into the nature and function of Treg cells in psoriasis in the natural state and during therapy will hopefully offer new and useful information about the pathogenesis of this difficult disease and open new avenues for treatment.

References

Akbar AN, Taams LS, Salmon M, Vukmanovic-Stejic M (2003) The peripheral generation of CD4$^+$CD25$^+$ regulatory T cells. Immunology 109:319–325

Annacker O, Pimenta-Araujo R, Burlen-Defranoux O, Bandeira A (2001) On the ontogeny and physiology of regulatory T cells. Immunol Rev 182:5–17

Aubin F, Mousson C (2004) Ultraviolet light-induced regulatory (suppressor) T cells: an approach for promoting induction of operational allograft tolerance? Transplantation 77:S29–S31

Austin LM, Coven TR, Bhardwaj N, Steinman R, Krueger JG (1999) The majority of epidermal T cells in psoriasis vulgaris lesions can produce type 1 cytokines, interferon-gamma, interleukin-2, and tumor necrosis factor-alpha, defining TC1 (cytotoxic T lymphocyte) and TH1 effector populations: a type 1 differentiation bias is also measured in circulating blood T cells in psoriatic patients. J Invest Dermatol 113:752–799

Baecher-Allan C, Brown JA, Freeman GJ, Hafler DA (2001) CD4$^+$CD25^{high} regulatory cells in human peripheral blood. J Immunol 167:1245–1253

Baecher-Allan C, Viglietta V, Hafler DA (2004) Human CD4$^+$CD25$^+$ regulatory T cells. Semin Immunol 16:89–98

Bata-Csorgo Z, Hammerberg C, Voorhees JJ, Cooper KD (1995) Kinetics and regulation of human keratinocyte stem cell growth in short-term primary ex vivo culture. Cooperative growth factors from psoriatic lesional T lymphocytes stimulate proliferation among involved, but not normal, stem keratinocytes. J Clin Invest 95:317–327

Bata-Csorgo Z, Cooper KD, Ting KM, Voorhees JJ, Hammerberg C (1998) Fibronectin and alpha5 integrin regulate keratinocyte cell cycling. A mechanism for increased fibronectin potentiation of T cell lymphokine-driven keratinocyte hyperproliferation in psoriasis. J Clin Invest 101:1509–1518

Bluestone JA, Tang Q (2004) Therapeutic vaccination using CD4$^+$CD25$^+$ antigen-specific regulatory T cells. PNAS 101:14622–14626

Braun-Falco O, Schmoekel C (1977) The dermal inflammatory reaction in initial psoriatic lesions. Arch Dermatol Res 258:9–16

Chatenoud L, Salomon B, Bluestone JA (2001) Suppressor T cells-they're back and critical for regulation of autoimmunity! Immunol Rev 182:149–163

Chen G, McCormick TS, Hammerberg C, Ryder-Diggs S, Stevens SR, Cooper KD (2001) Basal keratinocytes from uninvolved psoriatic skin exhibit accelerated and focal adhesion kinase responsiveness to fibronectin. J Invest Dermatol 117:1538–1545

Ellis CN, Krueger GG; Alefacept Clinical Study Group (2001) Treatment of chronic plaque psoriasis by selective targeting of memory effector T lymphocytes. N Eng J Med 345:248–255

Fehevari Z, Sakaguchi S (2004) Development and function of CD25$^+$CD4$^+$ regulatory T cells. Curr Op Immunol 16:203–208

Heng MC, Kloss SG, Kuehn SC, Chase DG (1986) Significance and pathogenesis of basal keratinocyte herniations in psoriasis. J Invest Dermatol 87:362–366

Hori S, Takahashi T, Sakaguchi S (2003) Control of autoimmunity by naturally arising regulatory CD4$^+$ T cells. Adv Immunol 81:331–371

Kriegel MA, Lohmann T, Gabler C, Blank N, Kalden JR, Lorenz HM (2004) Defective suppressor function of human CD4$^+$+ CD25$^+$ regulatory T cells in autoimmune polyglandular syndrome type II. J Exp Med 199:1285–1291

Krueger GG, Callis KP (2003) Development and use of alefacept to treat psoriasis. J Am Acad Dermatol 49:S87–S97

Krueger JG (2002) The immunologic basis for the treatment of psoriasis with new biologic agents. J Am Acad Dermatol 46:1–23

Lew W, Bowcock AM, Krueger JG (2004) Psoriasis vulgaris: cutaneous lymphoid tissue supports T-cell activation and 'Type 1' inflammatory gene expression. Trends Immunol 25:295–305

Morganroth GS, Chan LS, Weinstein GD, Voorhees JJ, Cooper KD (1991) Proliferating cells in psoriatic dermis are comprised primarily of T cells, endothelial cells, and factor XIIIa+ perivascular dendritic cells. J Invest Dermatol 96:333–340

Ramsdell F (2003) Foxp3 and natural regulatory T cells key to a cell lineage? Immunity 2003:165–168

Roncarolo MG, Levings MK (2000) The role of different subsets of T regulatory cells in controlling autoimmunity. Curr Op Immunol 12:676–683

Sugiyama H, Gyulai R, Toichi E, Garaczi E, Shimada S, Stevens SR, McCormick TS, Cooper KD (2005) Dysfunctional blood and target tissue CD4$^+$CD25^{high} regulatory T cells in psoriasis: mechanism underlying unrestrained pathogenic effector T cell proliferation. J Immunol 174:164–173

Szabo SK, Hammerberg C, Yoshida Y, Bata-Csorgo Z, Cooper KD (1998) Identification and quantification of interferon-γ producing T cells in psoriatic lesions: localization to both CD4$^+$ and CD8$^+$ subsets. J Invest Dermatol 111:1072–1078

Szeli M, Bata-Csorgo Z, Koreck A, Pivarcsi A, Polyanka H, Szeg C, Gaal M, Kemeny L (2004) Proliferating keratinocytes are putative sources of the psoriasis susceptibility-related EDA+ (extra domain A of fibronectin) oncofetal fibronectin. J Invest Dermatol 123:537–546

Shevach EM (2004) Regulatory/suppressor T cells in health and disease. Arth Rheum 50:2721–2724

Ting KM, Rothaupt D, McCormick TS, Hammerberg C, Chen G, Gilliam AC, Stevens SR, Culp L, Cooper KD (2000) Overexpression of the oncofetal

Fn variant containing the EDA splice-in segment in the dermal-epidermal junction of psoriatic uninvolved skin. J Invest Dermatol 114:706–711

Viglietta V, Baecher-Allan C, Weiner HL, Hafler DA (2004) Loss of functional suppression by CD4$^+$+CD25$^+$ regulatory T cells in patients with multiple sclerosis. J Exp Med 199:971–979

Vissers WH, Arndtz CH, Muys L, Van Erp PE, de Jong PM, van de Kerkhof PC (2004) Memory effector (CD45RO+) and cytotoxic (CD8$^+$) T cells appear early in the margin zone of spreading psoriatic lesions in contrast to cells expressing natural killer receptors, which appear late. Br J Dermatol 150:852–859

Ernst Schering Research Foundation Workshop

Editors: Günter Stock
Monika Lessl

Vol. 1 (1991): Bioscience ⇌ Socielty Workshop Report
Editors: D.J. Roy, B.E. Wynne, R.W. Old

Vol. 2 (1991): Round Table Discussion on Bioscience ⇌ Society
Editor: J.J. Cherfas

Vol. 3 (1991): Excitatory Amino Acids and Second Messenger Systems
Editors: V.I. Teichberg, L. Turski

Vol. 4 (1992): Spermatogenesis – Fertilization – Contraception
Editors: E. Nieschlag, U.-F. Habenicht

Vol. 5 (1992): Sex Steroids and the Cardiovascular System
Editors: P. Ramwell, G. Rubanyi, E. Schillinger

Vol. 6 (1993): Transgenic Animals as Model Systems for Human Diseases
Editors: E.F. Wagner, F. Theuring

Vol. 7 (1993): Basic Mechanisms Controlling Term and Preterm Birth
Editors: K. Chwalisz, R.E. Garfield

Vol. 8 (1994): Health Care 2010
Editors: C. Bezold, K. Knabner

Vol. 9 (1994): Sex Steroids and Bone
Editors: R. Ziegler, J. Pfeilschifter, M. Bräutigam

Vol. 10 (1994): Nongenotoxic Carcinogenesis
Editors: A. Cockburn, L. Smith

Vol. 11 (1994): Cell Culture in Pharmaceutical Research
Editors: N.E. Fusenig, H. Graf

Vol. 12 (1994): Interactions Between Adjuvants, Agrochemical
and Target Organisms
Editors: P.J. Holloway, R.T. Rees, D. Stock

Vol. 13 (1994): Assessment of the Use of Single Cytochrome
P450 Enzymes in Drug Research
Editors: M.R. Waterman, M. Hildebrand

Vol. 14 (1995): Apoptosis in Hormone-Dependent Cancers
Editors: M. Tenniswood, H. Michna

Vol. 15 (1995): Computer Aided Drug Design in Industrial Research
Editors: E.C. Herrmann, R. Franke

Vol. 16 (1995): Organ-Selective Actions of Steroid Hormones
Editors: D.T. Baird, G. Schütz, R. Krattenmacher

Vol. 17 (1996): Alzheimer's Disease
Editors: J.D. Turner, K. Beyreuther, F. Theuring

Vol. 18 (1997): The Endometrium as a Target for Contraception
Editors: H.M. Beier, M.J.K. Harper, K. Chwalisz

Vol. 19 (1997): EGF Receptor in Tumor Growth and Progression
Editors: R.B. Lichtner, R.N. Harkins

Vol. 20 (1997): Cellular Therapy
Editors: H. Wekerle, H. Graf, J.D. Turner

Vol. 21 (1997): Nitric Oxide, Cytochromes P 450,
and Sexual Steroid Hormones
Editors: J.R. Lancaster, J.F. Parkinson

Vol. 22 (1997): Impact of Molecular Biology
and New Technical Developments in Diagnostic Imaging
Editors: W. Semmler, M. Schwaiger

Vol. 23 (1998): Excitatory Amino Acids
Editors: P.H. Seeburg, I. Bresink, L. Turski

Vol. 24 (1998): Molecular Basis of Sex Hormone Receptor Function
Editors: H. Gronemeyer, U. Fuhrmann, K. Parczyk

Vol. 25 (1998): Novel Approaches to Treatment of Osteoporosis
Editors: R.G.G. Russell, T.M. Skerry, U. Kollenkirchen

Vol. 26 (1998): Recent Trends in Molecular Recognition
Editors: F. Diederich, H. Künzer

Vol. 27 (1998): Gene Therapy
Editors: R.E. Sobol, K.J. Scanlon, E. Nestaas, T. Strohmeyer

Vol. 28 (1999): Therapeutic Angiogenesis
Editors: J.A. Dormandy, W.P. Dole, G.M. Rubanyi

Vol. 29 (2000): Of Fish, Fly, Worm and Man
Editors: C. Nüsslein-Volhard, J. Krätzschmar

Vol. 30 (2000): Therapeutic Vaccination Therapy
Editors: P. Walden, W. Sterry, H. Hennekes

Vol. 31 (2000): Advances in Eicosanoid Research
Editors: C.N. Serhan, H.D. Perez

Vol. 32 (2000): The Role of Natural Products in Drug Discovery
Editors: J. Mulzer, R. Bohlmann

Vol. 33 (2001): Stem Cells from Cord Blood, In Utero Stem Cell Development, and Transplantation-Inclusive Gene Therapy
Editors: W. Holzgreve, M. Lessl

Vol. 34 (2001): Data Mining in Structural Biology
Editors: I. Schlichting, U. Egner

Vol. 35 (2002): Stem Cell Transplantation and Tissue Engineering
Editors: A. Haverich, H. Graf

Vol. 36 (2002): The Human Genome
Editors: A. Rosenthal, L. Vakalopoulou

Vol. 37 (2002): Pharmacokinetic Challenges in Drug Discovery
Editors: O. Pelkonen, A. Baumann, A. Reichel

Vol. 38 (2002): Bioinformatics and Genome Analysis
Editors: H.-W. Mewes, B. Weiss, H. Seidel

Vol. 39 (2002): Neuroinflammation – From Bench to Bedside
Editors: H. Kettenmann, G.A. Burton, U. Moenning

Vol. 40 (2002): Recent Advances in Glucocorticoid Receptor Action
Editors: A. Cato, H. Schaecke, K. Asadullah

Vol. 41 (2002): The Future of the Oocyte
Editors: J. Eppig, C. Hegele-Hartung

Vol. 42 (2003): Small Molecule-Protein Interaction
Editors: H. Waldmann, M. Koppitz

Vol. 43 (2003): Human Gene Therapy:
Present Opportunities and Future Trends
Editors: G.M. Rubanyi, S. Ylä-Herttuala

Vol. 44 (2004): Leucocyte Trafficking:
The Role of Fucosyltransferases and Selectins
Editors: A. Hamann, K. Asadullah, A. Schottelius

Vol. 45 (2004): Chemokine Roles in Immunoregulation and Disease
Editors: P.M. Murphy, R. Horuk

Vol. 46 (2004): New Molecular Mechanisms of Estrogen Action
and Their Impact on Future Perspectives in Estrogen Therapy
Editors: K.S. Korach, A. Hillisch, K.H. Fritzemeier

Vol. 47 (2004): Neuroinflammation in Stroke
Editors: U. Dirnagl, B. Elger

Vol. 48 (2004): From Morphological Imaging to Molecular Targeting
Editors: M. Schwaiger, L. Dinkelborg, H. Schweinfurth

Vol. 49 (2004): Molecular Imaging
Editors: A.A. Bogdanov, K. Licha

Vol. 50 (2005): Animal Models of T Cell-Mediated Skin Diseases
Editors: T. Zollner, H. Renz, K. Asadullah

Vol. 51 (2005): Biocombinatorial Approaches for Drug Finding
Editors: W. Wohlleben, T. Spellig, B. Müller-Tiemann

Vol. 52 (2005): New Mechanisms for Tissue-Selective Estrogen-Free
Contraception
Editors: H.B. Croxatto, R. Schürmann, U. Fuhrmann, I. Schellschmidt

Vol. 53 (2005): Opportunities and Challanges of the Therapies
Targeting CNS Regeneration
Editors: D. Perez, B. Mitrovic, A. Baron Van Evercooren

Vol. 54 (2005): The Promises and Challenges of Regenerative Medicine
Editors: J. Morser, S.I. Nishikawa

Vol. 55 (2006): Chronic Viral and Inflammatory Cardiomyopathy
Editors: H.-P. Schultheiss, J.-F. Kapp

Vol. 56 (2006): Cytokines as Potential Therapeutic Target
for Inflammatory Skin Diseases
Editors: R. Numerof, C.A. Dinarello, K. Asadullah

Supplement 1 (1994): Molecular and Cellular Endocrinology of the Testis
Editors: G. Verhoeven, U.-F. Habenicht

Supplement 2 (1997): Signal Transduction in Testicular Cells
Editors: V. Hansson, F.O. Levy, K. Taskén

Supplement 3 (1998): Testicular Function:
From Gene Expression to Genetic Manipulation
Editors: M. Stefanini, C. Boitani, M. Galdieri, R. Geremia,
F. Palombi

Supplement 4 (2000): Hormone Replacement Therapy
and Osteoporosis
Editors: J. Kato, H. Minaguchi, Y. Nishino

Supplement 5 (1999): Interferon: The Dawn of Recombinant
Protein Drugs
Editors: J. Lindenmann, W.D. Schleuning

Supplement 6 (2000): Testis, Epididymis and Technologies
in the Year 2000
Editors: B. Jégou, C. Pineau, J. Saez

Supplement 7 (2001): New Concepts in Pathology and
Treatment of Autoimmune Disorders
Editors: P. Pozzilli, C. Pozzilli, J.-F. Kapp

Supplement 8 (2001): New Pharmacological Approaches
to Reproductive Health and Healthy Ageing
Editors: W.-K. Raff, M.F. Fathalla, F. Saad

Supplement 9 (2002): Testicular Tangrams
Editors: F.F.G. Rommerts, K.J. Teerds

Supplement 10 (2002): Die Architektur des Lebens
Editors: G. Stock, M. Lessl

Supplement 11 (2005): Regenerative and Cell Therapy
Editors: A. Keating, K. Dicke, N. Gorin, R. Weber, H. Graf

Supplement 12 (2005): Von der Wahrnehmung zur Erkenntnis –
From Perception to Understanding
Editors: M. Lessl, J. Mittelstraß

This series will be available on request from
Ernst Schering Research Foundation, 13342 Berlin, Germany